Scaling: Why is animal size so important?

Scaling

Why is animal size so important?

KNUT SCHMIDT-NIELSEN
James B. Duke Professor of Physiology
Department of Zoology, Duke University

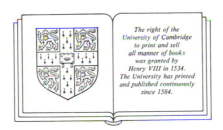

The right of the
University of Cambridge
to print and sell
all manner of books
was granted by
Henry VIII in 1534.
The University has printed
and published continuously
since 1584.

CAMBRIDGE UNIVERSITY PRESS

Cambridge

New York   Port Chester

Melbourne   Sydney

Published by the Press Syndicate of the University of Cambridge
The Pitt Building, Trumpington Street, Cambridge CB2 1RP
32 East 57th Street, New York, NY 10022, USA
10 Stamford Road, Oakleigh, Melbourne 3166, Australia

First published 1984
Reprinted 1985, 1986, 1989

Printed in the United States of America

Library of Congress in Publication Data
Schmidt-Nielsen, Knut, 1915-
   Scaling: Why is animal size so important?
   Includes index.
   1. Body size. 2. Morphology (Animals) I. Title.
QL799.S34   1984   596   84-5841
ISBN 0 521 2665 72 Hard covers
ISBN 0 521 3198 70 Paperback

Contents

Preface

This book is about the importance of animal size. Although it is informative and tells a great deal about what we know today, it is not encyclopedic. It should therefore be easy to read and use.

When we try to find the rules that govern animal function, we tend to think in terms of chemistry. We think of water, salts, proteins, enzymes, oxygen, energy, and so on – a whole world of chemistry. We should not forget that physical laws are equally important; they determine rates of diffusion and heat transfer, transfer of force and momentum, the strength of structures, the dynamics of locomotion, and so on. Physical laws provide possibilities and opportunities, yet they impose constraints and set limits to what is physically possible. It is our purpose to understand these rules because of their profound implications when we deal with organisms of widely different size and scale.

The book requires a minimum of basic numerical skills. It turns out that body size relationships are best expressed with logarithms and powers. These are simple enough to require familiarity with no more than a few easy algebraic operations. The entire book has been made as simple as possible, and most arguments are intuitively understandable. Although I use equations, which are useful for calculations, the conclusions are presented verbally, so that a minimum of effort is required to read the book from cover to cover.

Many facts pertaining to animal size call for rational explanations, and the reader will find that the book raises many questions. Amazing and puzzling information makes the book read a bit like a detective story, but the last chapter, which should give the final solution, does not answer all questions nor provide one great unifying principle. This presents a challenge that should stimulate further efforts, so that in the end we may better understand why the size of living things is of such fundamental importance.

1

The size of living things

Animals are different. An elephant differs from a mouse, both in shape and size. Size is one of the most important aspects of an animal's endowment, and yet size differences are so obvious that often we give no further thought to them. We know that the elephant is much bigger than the mouse, but we rarely think about how much bigger; in fact, an elephant weighs 100 000 times more than a mouse. The smallest shrew, when fully grown, is only one-tenth the size of a mouse, or one-millionth the size of an elephant.

The world we live in is governed by the laws of chemistry and physics, and animals must live within the bounds set by those laws. We shall see that body size has profound consequences for structure and function and that the size of an organism is of crucial importance to the question of how it manages to survive.

Let us take a closer look at the enormous size differences among living organisms (Table 1.1). Each single step in this table represents a 1000-fold difference in size, and the total difference between the smallest and largest organism listed is 10^{21}. The blue whale, which may exceed 100 tons, is the largest living animal, but the giant sequoia trees of California outweigh the largest whales by 10- or 15-fold.

Most of us cannot conceptualize what the exponential number 10^{21} really means. Assume that we estimate the size of a hypothetical giant organism larger than the blue whale by the same ratio, 10^{21}. This hypothetical super-giant organism would be 100 times the volume of the earth. The total mass of the universe will perhaps give a more convincing illustration of how difficult it is for us to grasp the magnitude of these large exponential numbers. Consider the incomprehensible magnitude of our universe, consisting of billions of galaxies with billions of stars in each; it is estimated to have a total mass of some 10^{80} gram.

Table 1.1. Size range of living organisms, arranged to show a 1000-fold difference in mass between each step.

Organism	Mass	
Mycoplasma	<0.1 pg	$<10^{-13}$ g
Average bacterium	0.1 ng	10^{-10} g
Tetrahymena (ciliate)	0.1 μg	10^{-7} g
Large amoeba	0.1 mg	10^{-4} g
Bee	100 mg	10^{-1} g
Hamster	100 g	10^2 g
Human	100 kg	10^5 g
Blue whale	>100 tons	$>10^8$ g

The smallest and the largest

The enormous size differences among organisms bring up two questions: (1) What are the consequences of a change in size? (2) Are there upper and lower limits to the size of living organisms?

Let us first look at the limits. The *Mycoplasma*, also known as PPLO (pleuropneumonialike organism), is the smallest organism we know that is able to live and reproduce by itself in an artificial medium. It is so small that, if the aqueous contents of the cell are at neutral pH, there will be, on the average, no more than two hydrogen ions inside the cell (Morowitz, 1966). Within this minute volume the cell must contain the necessary metabolic equipment of proteins and enzymes to be able to carry out its life processes; it must be able to grow and reproduce, and it must also carry the complete genetic information required for reproduction of the entire system. Because the macromolecules that carry the metabolic and genetic functions are essential, and their size probably cannot be reduced, the *Mycoplasma* cell may well represent an ultimate lower limit for the size of a living organism. Viruses, which are even smaller, have dispensed with the need for metabolic equipment; they consist essentially of only genetic information, parasitizing living cells that do the metabolic work of reproducing for them their genetic material.

At the other end of our size range is the 100-ton blue whale. It has been suggested, and more or less accepted as dogma, that the blue whale can reach its giant size only because it is aquatic, so that its enormous bulk is supported by water. Land mammals of a similar mass, it is said, would collapse under their own weight. The largest living land mammal is the 5-ton elephant. The fact that the much larger whales are aquatic

Figure 1.1. One of the large dinosaurs, the *Brontosaurus*, weighed around 30 tons. Other giant dinosaurs weighed perhaps three times as much. From Ostrom and McIntosh (1966).

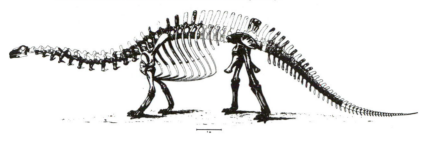

has been used to argue that animals much bigger than an elephant could not live on land; this, in turn, has been used to conclude that the extinct giant dinosaurs were too bulky to move on land and must have been semiaquatic.

Giant dinosaurs: Were they semiaquatic?

The largest dinosaur, *Brachiosaurus*, probably weighed about 80 tons (Colbert, 1962). This size may, in fact, be an underestimate, for the calculation was based on scale models and an assumed density of 0.9. The density of dinosaurs, like that of other vertebrates, was probably close to 1.0, and Colbert's estimate was perhaps too low by 10%. The longest dinosaur, *Diplodochus*, reached a length of 28.6 m, but was not as heavy. The long-necked *Brontosaurus* (Figure 1.1) was rather small, a mere 32 tons. Paleontologists have stated that "the long neck was an adaptation for life in deep waters" and that "this greatly simplified the problems of support and locomotion" (Romer, 1966). This hypothetical way of life is shown in Figure 1.2; the dinosaurs graze on underwater vegetation and use their long necks as a snorkel for breathing.

Can we believe this picture of a pleasant, semiaquatic life? One objection is that many fossil footprints of large dinosaurs show such a clarity of detail that they could hardly have been made under water (Gregory, 1951). Anyone who has tried to walk while immersed up to the neck in water knows that locomotion is not exactly simplified and that any footprints left in the mud are unlikely to show much detail. Also, breathing through a snorkel at a depth of 5 m requires the chest to work against the pressure of the surrounding water, which at that depth is 5000 kg/m^2. A human can barely breathe through a snorkel at a depth of 0.5 m, and at a much greater depth one's chest would be crushed by the water pressure.

Figure 1.2. The largest extinct dinosaurs were so heavy that it was once thought that they could not have moved freely on land and therefore must have led a semiaquatic life. Once submerged, as shown here, these animals would have had difficulty in moving about; furthermore, they would have had serious problems in breathing. From Gregory (1951) with courtesy of the Library Services Department, American Museum of Natural History.

The obvious conclusion is that the giant dinosaurs were not restricted to a semiaquatic life, and there is inadequate support for the contention that they were too heavy to move on land. Could vertebrates the size of dinosaurs be terrestrial? Or is terrestrial life a mechanical impossibility for giant vertebrates?

The largest land mammals

The largest land mammal that ever lived, so far as we know, was a relative of the modern rhinoceros, the herbivorous *Baluchitherium* from the Oligocene period (Figure 1.3). It stood over 5 m at the shoulder and weighed an estimated 30 tons, as much as the *Brontosaurus* (Granger and Gregory, 1935). Was this enormous animal too large to be safely supported by its skeleton? No paleontologist doubts that *Baluchitherium* was a fully terrestrial mammal, and because well-preserved skeletal material is available, we can estimate the strength of the bones from their size.

Three specimens of the metacarpal bones of *Baluchitherium* are shown in Figure 1.4, with the same bone from a modern rhinoceros

Figure 1.3. The largest land mammal that has ever lived, the *Baluchitherium,* was a relative of the modern rhinoceros. Its body weight has been estimated at about 30 tons. From Gregory (1951) with courtesy of the Library Services Department, American Museum of Natural History.

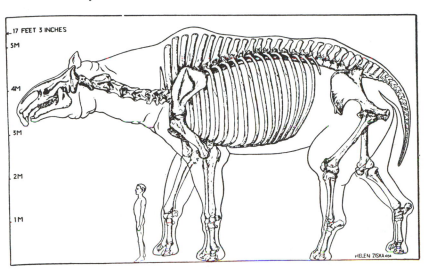

Figure 1.4. Metacarpal bones from three specimens of *Baluchitherium,* compared with the same bone from a modern rhinoceros (far left). The compressive strength of the largest metacarpal would have been about 280 tons, or nearly 10 times the body weight of the animal. From Gregory (1951) with courtesy of the Library Services Department, American Museum of Natural History.

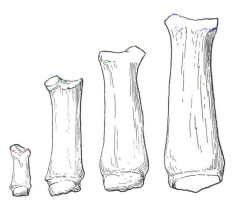

shown at the far left for comparison. The diameter of the largest meta-carpal is about 140 mm, and because the compressive strength of bone is about 1800 kg/cm^2 (Wainwright et al., 1976), we can calculate that in compression it could have supported about 280 tons. For a 30-ton animal, this gives a safety factor of nearly 10 times the body weight. It is probably no coincidence that this is roughly the same safety factor as that for static loads on the bones of humans. It is therefore reasonable to conclude that the strength of the skeleton of *Baluchitherium* was fully adequate for a terrestrial life.

We should realize, however, that the ability to support static loads is not the ultimate demand on the skeleton and therefore not the limiting factor in the size of terrestrial animals. Indeed, the support of static loads is probably irrelevant, for the stresses on the bones are much greater during locomotion when forces of acceleration and deceleration dominate and far exceed the forces of static loads. The fact that the active organism often operates close to its design limits is attested by the many pulled muscles, torn ligaments, and sprains suffered by human athletes in competitive sports.

Nevertheless, there is no reason to doubt that *Baluchitherium* was a plant-eating, land-living mammal, similar to the modern rhinoceros. It was one-third the size of the largest dinosaur, and it shows beyond doubt that the elephant by no means represents the size limit for terrestrial animals. Can we then conclude that even larger animals would be struc-turally sound? What is the ultimate size limit for land animals? Unfortu-nately, we have no conclusive answers to these questions.

Could the availability of food be a significant limitation on size? All the very large land mammals are plant eaters, and plant material is bulky and the digestion of cellulose is a slow process. The whales are better off in this regard: All the largest whales are filter feeders; they feed on plankton in a three-dimensional mass of water, and their food, mostly crustaceans, has a high energy value and is rapidly digested. We cannot say how important these factors are in the evolution of giant whales, for as in most evolutionary questions, we can only examine the available evidence and draw reasonable conclusions. Our conclusions, however, remain hypothetical, for we cannot experiment by building larger ele-phants, or giant whales that feed on cellulose. But we shall several times return to the question of what limits the size of animals.

2

Problems of size and scale

Definition of scaling

It is regrettable that we cannot study the effects of scaling by building super-sized elephants. Nevertheless, we can approach the problem in a different way, and in this regard we have much to learn from the engineer who continually solves the problems of building taller skyscrapers, longer bridges, bigger ships, and so on. Indeed, the need for changes in the size or scale of things has given rise to an entire branch of engineering known as scaling. For our purposes here, we shall use the definition that *scaling deals with the structural and functional consequences of changes in size or scale among otherwise similar organisms*.

If we increase the size of a brick house, we know that we need heavier foundations and thicker walls. There are practical limits to the size of brick houses, however, for the walls must be made thicker and thicker as house size increases. Eventually we meet an ultimate limit to further increases, dictated by the strength of brick. In the design of a skyscraper the engineer therefore changes the material in the main supporting structures; he uses steel rather than brick. In this case the constraint on a further increase in size is overcome by a change in material.

Another avenue is also open to the engineer: He can change to a new design. The construction of long bridges gives an example. By changing from the use of compression elements to tension elements for the main support, a 100-fold increase in the length of a bridge is quite feasible (Figure 2.1). Brick and stone are very strong in compression, but in tension they are weak and break easily. Steel, in contrast, has a high tensile strength, and relatively light elements suffice to support the elegant structures that the modern engineer uses to span kilometer-wide bodies of water.

Figure 2.1. When the length of a bridge is greatly increased, it is necessary to change its design. A bridge supported by a stone arch (bottom) has its main supporting elements in compression. A stone arch cannot be scaled up without risk of crumbling, but the design can be changed to a suspension bridge (top), in which the main supporting elements are steel cables in tension. Stone has high compressive strength but low tensile strength; steel, in contrast, has high tensile strength. From Schmidt–Nielsen (1975*a*).

Thus, three parameters can be changed when the size of a structure is to be increased:

1 Dimensions (e.g., thicker walls)
2 Materials (e.g., from brick to steel)
3 Design (e.g., from compression to tension elements)

Animals as well as engineers live in a physical world, and the same principles apply to both. The engineer can choose his dimensions, materials, and design; the biologist who looks at living organisms sees only the end products, but he wishes to understand the principles that make them living, functional organisms. On the one hand, what are the advantages of a certain size, material, and design? On the other, what are the limitations and constraints imposed by each of these three physical parameters?

What scale should the biologist use to measure the size of an organism? Two fundamental quantities can be measured with relative ease: mass and linear dimension. Mass is usually much to be preferred, but in some circumstances a linear measurement may be more convenient, more meaningful, and even more revealing. We shall come to such examples.

In most cases, however, the choice of a linear measurement causes problems: Which particular measurement is the most characteristic expression of the size of an organism? The problem is obvious. Say that we wish to compare a giraffe and a rhinoceros. How do we find a suitable linear measure to provide the necessary basis for valid comparisons? In contrast, weighing can be carried out with great accuracy, and structure and shape cause little difficulty, except for technical problems with animals the size of elephants and whales. Furthermore, mass is of fundamental importance in regard to such important matters as the strength of supporting structures (skeletons), the demands on the muscular system in locomotion, the need for food, and so on. Usually, mass is also an adequate measure of volume, for nearly all animals have densities close to 1.0. However, within limits, characteristic linear dimensions may provide a suitable or useful measure, as we shall later see in the discussion of problems pertaining to swimming fish.

Constraints can be overcome by a novel design

Some examples will make it obvious that there are inescapable biological consequences of size and design. For example, swimming with the aid of cilia or flagella is possible only for very small organisms, and fishes use a different propulsive mechanism. A paramecium covered with cilia swims many times its body length in a second, but a giant shark covered with cilia would get nowhere. The laws of fluid dynamics can, in a more formal way, explain why microorganisms and fish, from this point of view, seem to live in different worlds.

Another difference between the very small organism and the large organism that is intuitively understandable concerns the supply of oxygen. Diffusion over short distances is very fast, and oxygen can rapidly reach all parts of a microorganism. Over longer distances, diffusion is a slow process, and diffusion alone is totally inadequate for supplying oxygen to a large animal. A novel principle, transport by convection, is then added to augment the inadequate diffusion process.

Convection is important both outside and inside the organism. Convection in the external medium is called ventilation, whether oxygen is obtained from water or air and whether the animal has gills or lungs. Convection of fluids within the body is called circulation. Thus, the new principle, mass transport by convection, becomes essential as the size of an animal increases.

Pumping of fluids, whether external or internal, requires energy. Another novel principle can help reduce the work of pumping the internal fluid. If a respiratory pigment, such as hemoglobin or hemocyanin, is added to the fluid, its oxygen-carrying capacity can be increased perhaps 100-fold. The corresponding reduction in the volume required reduces the energy demand on the pumping mechanism by two orders of magnitude. Thus, a new design has again removed certain constraints, and therefore potential limits, to the size of organisms. We full well know that without hemoglobin and circulation, large active mammals as we know them would not be possible.

There are, however, other options for rapid internal transport of oxygen, and in this regard insects have gone their own separate way. They do not use convection in a fluid (blood) to augment gas transport, for their respiratory system (the tracheae) is based on diffusion in gas rather than in water. The tracheal system consists of tubes that extend throughout the body, and the fact that the diffusion coefficient for oxygen in air is some 10 000 times higher than in water ensures an adequate supply of oxygen without the aid of the blood. Nevertheless, many highly active insects use convection to help speed up the ventilation of their respiratory systems by active pumping of air through the major branches of the tracheal system. Even an insect as small as the fruit fly (*Drosophila*) uses active ventilation of the thorax to supply its wing muscles with oxygen at the high rate required during flight (Weis–Fogh, 1964).

Because insect blood does not help in the transport of oxygen, and because diffusion slows down with increasing distance, does the respiratory system in insects impose a limit on their body size? Is this the reason that all insects are small? Or could a hypothetical super-sized insect supply its most distant organ with sufficient oxygen? Other arthropods, notably the largest crabs, may be 1000 times larger than the largest insects; crabs breathe with gills, and their blood contains the oxygen-carrying pigment hemocyanin. However, from what we know, it appears that the dimensions of the tracheal system are ample for gas transport, for the larger tracheae can be actively ventilated, reducing the distances over which diffusion must take place.

Perhaps other design constraints limit the body size of insects? It seems that a limitation may reside in their structure, notably in the characteristics of the exoskeleton. Although we do not fully understand the principles of skeletal design, some consequences and limitations of exoskeletons have been clearly set out by Currey (1967). An exoskeleton

Figure 2.2. In isometric triangles, or in other isometric figures, all corresponding linear dimensions have the same relative proportions.

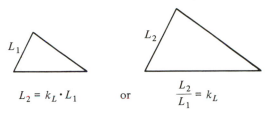

$$L_2 = k_L \cdot L_1 \qquad \text{or} \qquad \frac{L_2}{L_1} = k_L$$

provides both armor and support, and the requirements for support differ in air and in water. All the really large arthropods are aquatic, the record probably being the Japanese spider crab, whose legs may span 4 m (Schmitt, 1965). We shall later discuss the mechanical properties of skeletons and the advantages and constraints that pertain to body size.

Similarity

The idea of similarity first acquired a precise meaning in geometry more than 2000 years ago. In Euclidean geometry, two triangles are similar if corresponding sides are in a constant ratio and corresponding angles are equal (Figure 2.2). Euclid's fourth proposition states that if two triangles are equiangular, then their corresponding sides are proportional. The converse is also true: If corresponding sides are in equal proportions, then corresponding angles are equal. We say that two such triangles are *geometrically similar.*

In two geometrically similar triangles (Figure 2.2), two corresponding sides, L_1 and L_2, are related as follows:

$$L_2 = k_L L_1 \quad \text{or} \quad \frac{L_2}{L_1} = k_L$$

The other corresponding sides will then be related in the same ratio, k_L, which can be called the *similarity ratio.* The same holds true of any other corresponding linear measurements, such as the heights of the triangles; all such corresponding linear measurements will be in the ratio k_L.

The same considerations pertain to any other geometrically similar figures and can be extended to three-dimensional figures as well. Any corresponding linear dimensions on two geometrically similar bodies, whether cubes, pyramids, cones, or more complex shapes, will be in the same constant proportion.

Figure 2.3. In geometrically similar (isometric) bodies, all corresponding linear dimensions are related in the same proportion, and all corresponding surfaces have areas that are related in the same proportion squared.

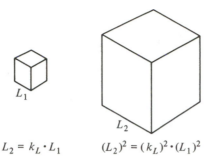

$$L_2 = k_L \cdot L_1 \qquad (L_2)^2 = (k_L)^2 \cdot (L_1)^2$$

Geometrically similar bodies are often called *isometric,* and for convenience we shall use the two terms interchangeably, often with a preference for isometric because it is shorter. *Isometric then stands for the well-defined concept of geometric similarity.*

Now consider two cubes of different sizes (Figure 2.3). Because all corresponding linear measurements of the two cubes are in the same proportion and all corresponding angles are equal, the two cubes are geometrically similar or isometric.

The surface areas of the two cubes, however, do not change in the same ratio as their linear dimensions, but rather with the square of the linear ratio. We could write

$$L_2 = k_L \times L_1$$
$$(L_2)^2 = (k_L)^2 \times (L_1)^2$$

Similarly, the volumes of the cubes change in proportion to the third power of their linear dimensions. Say that the larger cube has a side twice as long as the smaller cube. Its surface area will then be 2^2, or four times that of the smaller cube, and its volume will be 2^3, or eight times as great.

The same rules apply to any other geometrically similar or isometric three-dimensional bodies, whatever the shapes. Therefore, the rule applies also to objects as irregularly shaped as animals; if two dogs of different size are indeed isometric, their surfaces and volumes will be in ratios related to their linear dimensions to the second and third power, respectively.

We can write out these essentials of isometric geometry as follows:

Figure 2.4. If the surface area of a cube is plotted against the volume of the cube, the relationship is nonlinear. That is, the surface area does not increase in proportion to the volume of the cube, but becomes smaller in relation to the volume for larger cubes.

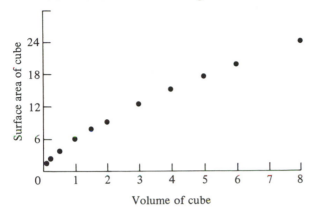

$$\text{surface} \propto (\text{length})^2 \quad \text{or} \quad S \propto L^2 \tag{1}$$

$$\text{volume} \propto (\text{length})^3 \quad \text{or} \quad V \propto L^3 \tag{2}$$

$$\text{surface} \propto (\text{volume})^{2/3} \quad \text{or} \quad S \propto V^{2/3} \tag{3}$$

The last equation simply states that as the volume of a body is increased, its surface does not increase in the same proportion, but only in proportion to the two-thirds power of the volume, a well-known yet important point to remember.

Let us express this simple fact in the form of a graph. For simplicity, consider the surface areas of cubes of various volumes (Figure 2.4). The curve we see corresponds to the equation $S = 6V^{2/3}$ and merely repeats the verbal statement that the surface of a cube increases less rapidly than its volume.

Should we choose to plot the surface area of the cube against its volume on logarithmic coordinates, however, we obtain a different graph (Figure 2.5). The fully drawn straight line corresponds to the logarithmic form of the preceding equation: $\log S = \log 6 + \frac{2}{3} \log V$. For bodies of other shapes, the factor 6 will change, but for isometric scaling the exponent $\frac{2}{3}$ will remain constant; that is, the surfaces of any two isometric bodies are related to their volumes by the power 0.67.

Let us now remember that the surface of isometric bodies, when plotted against their volume on logarithmic coordinates, is represented

Figure 2.5. If the surface area of a cube is plotted against the volume on logarithmic coordinates, we obtain a straight regression line with a slope of 0.67. If, instead, the surface area *per unit volume* of the cube is plotted (dashed line), the regression line shows that the relative surface area decreases with increasing size of the cube. The slope of the dashed line is −0.33.

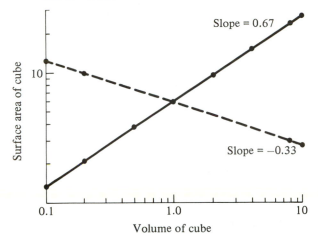

by the straight line with the slope 0.67 (the slope being defined as $\Delta y/\Delta x$).

It may be useful to repeat another well-known fact: that smaller bodies have, relative to their volumes, larger surface areas than larger bodies of the same shape. This can be expressed by dividing equation (3) by V:

$$S = k \times V^{0.67}$$

$$\frac{S}{V} = k\frac{V^{0.67}}{V} = k \times V^{0.67 - 1.0}$$

$$\frac{S}{V} = k V^{-0.33} \tag{4}$$

Equation (4) is the expression for the surface area per unit volume for any geometrically similar bodies. Plotted on logarithmic coordinates, it will give a straight line with the slope −0.33. For cubes, the dashed line with a negative slope in Figure 2.5 shows how the relative surface area of a cube decreases as its volume increases.

Allometric scaling

Real organisms usually are not isometric, even when organized on similar patterns. Instead, certain proportions change in a regular fashion, and many such examples will be mentioned in the following sections. In biology, such *nonisometric* scaling is often referred to as *allometric* (from the Greek *alloios*, which means different). An amazing number of morphological and physiological variables are scaled, relative to body size, according to allometric equations of the general form

$$y = a \cdot x^b$$

$$\log y = \log a + b \log x$$

This equation expresses the simple statement, thoroughly familiar to biologists, that when two variables are plotted on logarithmic coordinates, the result is a straight line. A great variety of observations that relate biological variables to body size conform to this general equation, in which the exponent b represents the slope of the straight line obtained in a logarithmic plot.[1]

The exponent (slope) can take on different values and can be either positive or negative, depending on the function being considered. When we buy potatoes or apples, the amount we pay increases in proportion to the amount we buy. This is a simple proportionality, and a plot will show a straight-line relationship with a slope of 1.0 (Figure 2.6A). The same relationship applies to the blood volume in mammals; the blood makes up a constant fraction of the body mass, and the larger the animal, the more blood.

As we saw earlier, the skeleton of a large animal is relatively heavier than that of a small animal. With increasing size, the skeleton increases out of proportion to the increase in body mass (Figure 2.6B). This is reflected in the slope, which for this function is greater than 1.0.

If the dependent variable increases at a slower rate than would be indicated by simple proportionality, the regression line will have a slope less than 1.0. A well-known example is metabolic rate, which increases with body size, but less than would be indicated by simple proportionality. In this particular case the slope of the regression line is 0.75 (Figure 2.6C).

If we consider a quantity that does not change with body size, the slope will be zero (Figure 2.6D). For example, hemoglobin concentration in the blood is similar in all mammals. There are some variations from the mean, but the variations we may find will not be related to body size.

---

1 Various authors have used an array of different symbols in allometric equations. For example, the exponent has been noted as a, b, e, k, n, s, x, α, β, and γ. Some of these symbols are meaningless, for the exponent is neither a constant (k) nor an independent variable (x). In current literature, the letter b is usually used for the exponent, and that notation will be used throughout this book. In my earlier writings on this subject I used the exponent a (Schmidt–Nielsen, 1970, 1975a), but the common practice is now to use b.

Figure 2.6. The regression lines for different exponents in the equation $y = ax^b$ (dashed lines) have different slopes depending on the value of b. The slope (b) may indicate proportionality (fully drawn line), but often deviates in a regular way from proportionality. All plots refer to logarithmic coordinates.

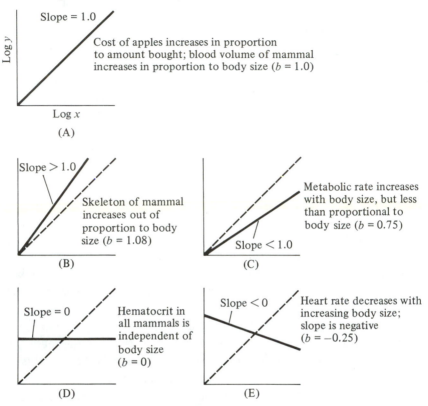

We may also find functions that decrease with increasing body size, which will give a negative slope to the regression line. This is the case with heart rate in mammals. An elephant may have a heart rate of perhaps 25 or 30 beats per minute, whereas a mouse has a rate of several hundred. For mammals in general, heart rate decreases with body size, and the regression line has a negative slope of -0.25 (Figure 2.6E).

All the plots in Figure 2.6 are on logarithmic coordinates, and the regression lines are straight lines (as they should be when they represent logarithmic functions). It is useful to compare two plots of the same function, one on arithmetic coordinates and one on logarithmic coordinates. Figure 2.7A shows functions with exponents of 1.0 and 0.75. Only the slope of 1.0 yields a straight regression line, and the difference

Figure 2.7. Logarithmic functions that on linear coordinates yield strikingly different regression lines (A) give straight lines that appear to have closer similarity when plotted on logarithmic coordinates (B). It should be noted that an apparently minor difference in the slopes of two regression lines on logarithmic coordinates can be numerically very important.

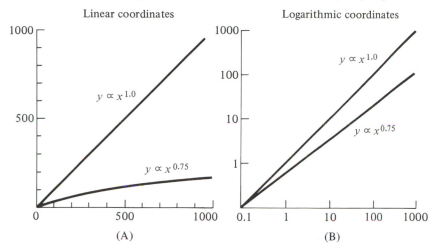

between the lines is striking. When the same functions are plotted on logarithmic coordinates, both give straight lines (Figure 2.7B), and the lines do not give the appearance of being very different. This should be kept carefully in mind; what appear to be small differences in exponents represent sizable magnitudes when expressed arithmetically.

Dimensions

The system of Newtonian mechanics operates with the fundamental quantities mass, length, and time. These can be represented by the dimensional symbols M, L, and T. It is important to note that these symbols do not stand for real numbers. Thus, the dimensional symbol L designates the dimension length, and no number or unit can be attached to it. A given distance, on the other hand, is a physical quantity and is described by a numerical value and a unit.

The dimensional symbols form a simple but consistent mathematical language. The symbols can be manipulated according to certain rules, the most important being that they can be used only for multiplication and division, not for addition or subtraction. An important use for dimensional symbols is in manipulations known as dimensional analysis, a discipline that is of great importance in engineering but thus far has been little used in biology.

In this context, the following quotation from the physicist P. W. Bridgman is worth remembering: "It is to be especially noticed that the results of dimensional analysis cannot meaningfully be applied to any system whose fundamental laws have not yet been formulated in a form independent of the size of the fundamental units. For instance, dimensional analysis would certainly not apply to the results of many biological measurements, although these may perfectly well have physical validity as descriptions of the phenomena. At the present time, it seems that biological phenomena can be described in complete equations only with the aid of as many dimensional constants as there are physical variables. In this case, we have seen, dimensional analysis has no information to give" (Bridgman, 1937, p. 53).

One essential rule in the use of dimensions is that equations, in order to be meaningful, must be dimensionally consistent or homogeneous. This rule was recognized by Fourier more than 150 years ago, but its important consequences are often overlooked.

Let us examine how this works for the well-known Poiseuille equation, which describes the flow of fluids through a cylindrical tube:

$$\text{flow rate} = \frac{\pi \rho P r^4}{8 \eta l}$$

The symbols used and their dimensions are as follows (see Appendix E):

ρ = density of fluid $\qquad$ $M L^{-3}$

P = pressure difference $\qquad$ $M L^{-1} T^{-2}$

r = radius of tube $\qquad$ L

η = viscosity of fluid $\qquad$ $M L^{-1} T^{-1}$

l = length of tube $\qquad$ L

Insertion of the dimensions for each variable gives the following expression:

$$\frac{(M L^{-3}) \times (M L^{-1} T^{-2}) \times (L^4)}{(M L^{-1} T^{-1}) \times (L)} = M T^{-1}$$

The dimensions of Poiseuille's equation thus are reduced to mass over time; that is, the flow rate is expressed as mass of fluid per unit time. This tells us that the equation is dimensionally homogeneous and that the two constants in the equation, π and 8, are therefore dimensionless, and furthermore, that the equation therefore is valid for any consistent system of units.

Sometimes Poiseuille's equation is given as

$$\text{flow rate} = \frac{\pi P r^4}{8 \eta l}$$

This differs from the form we used earlier, which we found to be dimensionally correct. Is this second form wrong? A quick test of dimensions gives the following result:

$$\frac{(M L^{-1} T^{-2}) \times (L^4)}{(M L^{-1} T^{-1}) \times (L)} = L^3 T^{-1}$$

We can now see that this equation dimensionally yields volume (L^3) per unit time, which is flow rate. A comparison of the two equations shows that the term for density of the fluid, ρ, is missing from the second. Our examination of dimensions shows both equations to be dimensionally correct and rapidly reveals the difference.[2]

Dimensionless quantities

Some quantities are known as dimensionless quantities. The most obvious dimensionless quantity is the ratio of two physical quantities of the same kind. Say that we determine the ratio between the long side and the short side of this page and find it to be 1.5. This number is a ratio of two quantities of the same dimension, length, and that gives a dimensionless ratio. The ratio is independent of the choice of units, whether millimeters, centimeters, or inches (but these must, of course, be consistent).

Other examples of dimensionless quantities include strain and the coefficient of friction. Strain is the ratio of a change to the total value of the quantity in which the change has occurred. Similarly, the coefficient of friction is the ratio of the force required to move one surface over another to the total force pressing the surfaces together. Both are dimensionless ratios.[3]

A dimensionless ratio frequently used in fluid dynamics is the Reynolds number, which is a ratio between inertial and viscous forces. The importance of the Reynolds number will be explained in later chapters and will not be discussed here.

2 It should be noted that there is nothing sacred about the dimensions M, L, and T. In engineering, force is extensively used as a fundamental dimension, together with length and time, giving an F L T system. In this system, force = F, mass = $F L^{-1} T^2$, work = F L, power = $F L T^{-1}$, and so on. This can be compared with the dimensions in the M L T system listed in Appendix E.

3 Dimensionless ratios of the kind described here have the same mathematical property as dimensional symbols: They can be multiplied or divided, but they cannot be added or subtracted.

It may be useful, however, to discuss briefly a few dimensionless ratios that are descriptive of the mammalian body. Let us consider an example. Observations have shown that, in general, mammals have hearts that, with a certain leeway for variation, have a mass that is 0.6% of the body mass. Similarly, mammals generally have an amount of blood that is about 5% of their body mass. Let us write these two statements in the following equations:

$$M_{heart} : M_{body} = 0.006$$

$$M_{blood} : M_{body} = 0.05$$

Because the two quantities, heart mass and blood mass, are both related to body mass, they must be related to each other, and we can write

$$\frac{M_{blood}}{M_{heart}} = \frac{0.05}{0.006} = 8.3$$

This ratio states that the mass of the blood (or, for practical purposes, its volume) in a mammal is roughly eight times the mass of the heart (or its volume). This statement (from which there may be deviations) is a description of blood volume and heart size in mammals that, in a general way, applies to all body sizes, be it a mouse or an elephant.

In the next chapter we shall consider more examples of allometric equations and how they can be used as simple tools to express size relationships.

3

The use of allometry

Biological significance and statistical significance

Properly calculated allometric equations (or regression lines) will be accompanied by statistics that give information about significance and confidence limits. Statistics are necessary because we cannot rely on subjective evaluations of whether or not data and numbers are significant.

Allometric equations, $y = ax^b$ (or corresponding linear-regression lines), have two important numerical terms: the *proportionality coefficient a* (the intercept at unity) and the *exponent b* (the slope of the regression line). These two terms have different meanings and can answer different questions. An example will help.

The proportionality coefficient can be used to answer questions such as this: Do marsupials, in general, have lower metabolic rates than eutherian mammals (see p. 64)? The equations for the metabolic rates of these two groups have the same exponent, and we can therefore directly compare the proportionality coefficients, which are lower for marsupials. This tells us that marsupials, in general, have lower metabolic rates than eutherian mammals. The exponent, on the other hand, tells us that the metabolic rate changes with changing body size in the same way in marsupials and eutherian mammals. This suggests that the same principles determine the scaling of metabolic rates in the two groups (although the coefficient told us that the levels of their metabolic rates differ systematically).

In this example we probably feel confident that our conclusions are valid. However, to decide whether a difference between two numbers is insignificant or carries a meaning, we need statistical information. Depending on the use we want to make of the equations, the needed statistics will vary, but we would always like to know the confidence limits for the coefficient *a* and the exponent *b*, which are more informative than correlation coefficients. However, after having established the statistical significance, we run into trouble unless we keep in mind the essential difference between statistical and biological significance.

Assume that we have put together an equation or a regression line with proper statistics, correlation coefficient, standard deviation, confidence limits, and so on. We must realize that *the proper statistics are no more than a description of the numbers at hand.* An evaluation of the biological significance necessitates consideration of how the numbers were obtained, and two important factors enter here: One pertains to measurement errors, which may be random or systematic, and the other to how our sample was obtained, for example, the animal species we could get our hands on. Measurement errors may extend from simple measurement errors to all sorts of mistakes, and sampling often involves those animals we happen to have in our laboratory or can most readily obtain. In the early studies by Kleiber and Brody, who related metabolic rates of mammals to their body sizes, the large animals were cows, horses, swine, goats, and sheep, that is, only domesticated animals, and these may or may not be representative of mammals in general. There could easily be a systematic bias in the sample that no amount of statistical treatment could remove. An increased number of measurements to help increase the statistical significance of a sample and narrow the confidence limits does not necessarily aid in improving biological significance.

This is important: Highly significant statistics do not signify equally high biological significance.

The allometric signal

Let us use a graph from later in this book: Figure 6.1. The main message carried by the regression line is called the *signal.* In this case the signal is that the rate of oxygen consumption in mammals increases with body size to the power 0.75. This allometric signal is contained in the information at hand and can equally well be expressed in a graph or a logarithmic equation.

We know that no single observation will fall exactly on the regression line and that there will be smaller and larger deviations. The line represents the statistical best fit, provided we have decided to calculate a straight regression line on logarithmic coordinates. This line is then calculated on the basis of the log-transformed original data, which is the customary way of handling scaling problems in biological material. There are valid theoretical reasons for this practice, and those who are interested in the validity of this procedure may wish to consult the work of Zar (1968), Lasiewski and Dawson (1969), Smith (1980), Zerbe et al. (1982), and others.

When we calculate a linear-regression line, we have already made some decisions, and we should say the following: "*If* we decide to calculate a linear-regression line on logarithmic coordinates, this is what we obtain, and here are the appropriate statistics." This is important, because the allometric equations or linear-regression lines are no more than descriptions of a set of numbers at hand.

Secondary signals

Assume that an observation deviates substantially from what we consider a reliable estimate of a regression. This deviation may raise a question of whether or not this observation belongs within the category or population we believe we are dealing with. Again, an example may help. Iversen (1972) reported that weasels have metabolic rates about twice as high as would be expected for ordinary mammals of their body size. Assume that there are no errors in the determinations, calculations, and so on, and that we have a consistent deviation from the allometric equation for mammals. This deviation is then a *secondary signal,* saying that weasels deviate from the general allometric signal that describes the metabolic rates of mammals in general. Such a deviation is no surprise to a biologist familiar with the nervous activity of weasels, just as he would not be surprised to find that a sloth has a lower rate than the "expected" metabolic rate.

Not only must we decide whether a given deviation is random or a secondary signal, we also come to the difficult question whether or not such deviants should be included in the calculation of a general regression. Presumably, if enough different animals are included in our sample, deviations above and below will not grossly distort the signal contained in the general description of the allometric relationship. However, if we have a large number of data points on one side of the regression line and few on the other, distortion is inevitable. There is probably no objectively valid way of handling this difficult problem.

Outliers and extrapolations

We now come to an important but often disregarded consideration: the role of points at the extremes of the range of data, the outliers. The customary calculation of regression lines is based on the method of least squares. Our samples usually have the largest population in the

Figure 3.1. A regression line calculated by the method of least squares will have a slope disproportionately more affected by an outlier (y') than by a point near the middle. The confidence limits (dotted lines) for the regression become increasingly wider toward the upper and lower extremes of the range.

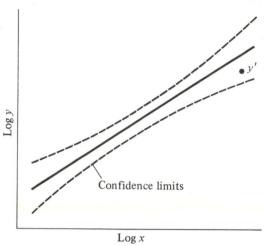

middle, with fewer outlying points. The outliers, however, have a disproportionate effect on the calculation of a regression line.

The calculated regression line will always pass through the point representing the mean of the x values and the mean of the y values. Now consider an outlier, a single point, y', to the far right (Figure 3.1). This point, because of its location, affects the regression line unduly because its deviation is squared for the calculation. A point near the mean value, in contrast, has little or no effect on the slope. Thus, outliers have a disproportionate effect on the slope of a regression line, and any sampling error or bias is particularly important for the outliers.

In Brody's calculation of the allometric equation for metabolic rates in mammals, the largest animal was the elephant. Brody decided that the metabolic rate for elephants should be adjusted, because these animals could not be measured while fasting, and he used a value with "30% deducted from the original value (10% for standing & 20% for heat increment of feeding)" (Brody, 1945, p. 389). The validity of this adjustment can be questioned, because it produces an equation not based on the actual observations. However, whether or not the "correction" was justified, such a subjective decision pertaining to an outlier has a disproportionate effect on the end result.

There is another inherent weakness pertaining to outlying points. If an allometric equation or regression line is used to evaluate whether or not a given observation is consistent with an established regression equation, we must ask whether the observation in question falls within or outside the confidence limits of the regression line. The simplest way of making the point for the outliers is shown in Figure 3.1. The combined effect of the confidence limits on the proportionality coefficient and on the exponent is that the confidence limits on the regression are increasingly wide toward the two ends of the regression line.

The consequence is paradoxical. An outlier has an unduly great effect on the location of a regression line, and this is where we can be the least confident of the significance of a deviant. This leads us to another important matter: extrapolation.

First of all, a calculated regression line or equation is valid only in the range of the data at hand. We have seen that the confidence limits deteriorate toward the extremes and will continue to do so beyond the range on which the regression is based. This increasing uncertainty is in itself bad enough, but we have no way of knowing what happens beyond the observed data.

The most important objection to extrapolations is our ignorance of constraints or discontinuities that may apply beyond the range in which we have observations. Unless we fully understand all the pertinent factors involved (and that is not likely to happen in a biological system), going beyond the limits of observation is not simply chancy, but outright perilous.

The use of allometric equations

There are probably good theoretical reasons for the widespread use of allometric equations in biology, but even if these reasons are inadequately understood, the empirical use of such equations is very helpful. Some examples should help clarify the use and the convenience of allometric equations as they pertain to biological structure and function.

As far as I know, the first use of an allometric equation to express a biological relationship was that by Snell (1891). Snell was interested in a method of comparing the mental capabilities of various mammals in relation to their brain size. However, the brain makes up a smaller fraction of the total body mass of a larger mammal, and Snell wanted to take this into account. For that purpose he developed an equation to express the mass of the brain in mammals that in our notation would read

$$M_{\text{brain}} = a M_{\text{body}}^{b}$$

The exponent (designated "somatic exponent" by Snell and given the symbol S) was assumed to be the same for all mammals. It was said to be very close to 0.68 (i.e., the brain would increase with body size for mammals in almost exact proportion to the body surface). We shall see that this century-old figure probably is not in need of much revision, for recent compilations (Stahl, 1965) indicate that the mass of the brain relative to body mass for mammals (in kg) is well described by the equation

$$M_{\text{brain}} = 0.01 \, M_{\text{body}}^{0.70}$$

Furthermore, the exact value of the exponent varies somewhat, depending on the data base available to different authors.

Were dinosaurs stupid?

Allometric equations, or the equivalent graphic regression lines, are convenient and useful tools in biology. Consider the poorly documented but often repeated statement that the giant dinosaurs had very small brains relative to their body size and that competition from the smaller but brainier mammals led to their ultimate extinction.

A careful examination of this statement has been made by Jerison (1969, 1970). His compilation of brain sizes for various vertebrates shows that reptiles and fish have smaller brains than mammals and birds, without any overlap between the groups. By making casts of the brain cavities in fossil dinosaur skulls, Jerison obtained objective information about their brain sizes. Dinosaurs did indeed have smaller brains than modern mammals, but even so, their brain size was perfectly within the range that could be expected for reptiles of their body size (Figure 3.2). Because modern reptiles, with their relatively small brains, have survived competition with mammals for hundreds of millions of years, it is difficult to accept as a conclusive argument for dinosaur extinction that they had disproportionately small brains.

The use of graphs is a convenient way of presenting material, but the corresponding allometric equations often provide a more convenient basis for comparisons. Instead of the range of mammalian brain sizes indicated in Figure 3.2, we can use the equations that represent the best-fitting regression lines for the available data (Table 3.1).

The first line in Table 3.1 shows that a typical mammal of 1 kg body mass can be expected to have a brain size of 0.01 kg and that the brain size increases in proportion to the body mass to the power 0.70. The

Figure 3.2. Brain sizes for various vertebrates fall within ranges characteristic for each group. Within each group, brain size increases roughly with body size to the $\frac{2}{3}$ power. The brain sizes of the large extinct dinosaurs fall within an extension of the range characteristic for modern reptiles, and the statement that dinosaurs had disproportionately small brains does not seem justified. From Jerison (1970).

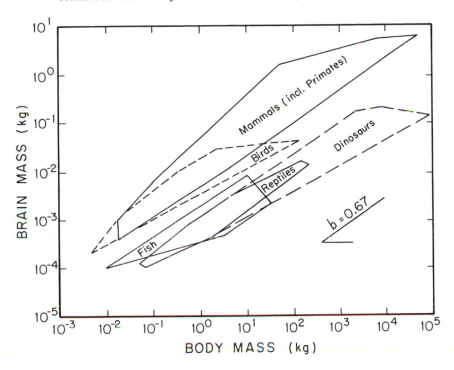

Table 3.1. Average brain size for mammals as a group and for major primate groups, expressed as a function of body mass (M_b, in kg); data from Stahl (1965).

Animal group	Brain size
Mammals	$0.01\,M_b^{0.70}$
Monkeys	$0.02\text{–}0.03\,M_b^{0.66}$
Great apes	$0.03\text{–}0.04\,M_b^{0.66}$
Humans	$0.08\text{–}0.09\,M_b^{0.66}$

following lines in Table 3.1 show equations for brain sizes for the major groups of primates, which all vary with nearly the same power of body mass, 0.66.

When the numerical values of the exponents in the equations are the same, the proportionality coefficients preceding the exponential term can be used directly to compare the magnitude of the variable in question, in this case brain size. These coefficients show that monkeys, in general, have brains two to three times as large as typical mammals, that great apes have brains twice as large as monkeys, and that humans have brains twice as large again. This brief analysis of these equations expresses the essential difference in brain size between monkey and human, and the important fact is that we can make this comparison directly, although there is no overlap in body size between these two groups. Thus, this simple allometric comparison of brain sizes is analogous to the graphic comparison of dinosaur brains and other reptile brains shown in Figure 3.2.

This approach to the study of scale effects and body size is a valuable tool, but it is also a powerful device that can be abused. This may have been the case when allometric scaling was used to estimate the wingspan of a fossil flying reptile, a pterosaur, not unlike the *Pteranodon* shown in Figure 3.3. The only wing bone available for measurement was the bone in the upper arm, the humerus. It was used to estimate, by extrapolation, a total wingspan for this animal of 15.5 m, "undoubtedly the largest flying creature presently known" (Lawson, 1975). The extrapolation was based on the proportion between the short humerus and the total wing length in other pterosaurs of smaller body size from which wing bones have been better preserved. The bone on which the extrapolation was based was 0.52 m long, or roughly 1/30 of the estimated wingspan. Extrapolating from a bone that is only 3% of the calculated wingspan is roughly comparable to estimating the armspan of a man from the length of his thumb.

True enough, the enormous wingspan was criticized by Greenewalt (1975a), who suggested that such a calculation could more realistically be based on a comparison involving equations for observed wing lengths of birds. By using the relationship between the length of the humerus and the wingspan for 139 species of birds, Greenewalt suggested that the fossil pterosaur might have had a wingspan only one-third of the estimated 15 m, or 5.25 m. This sounds more realistic, but it is difficult to say whether or not it is a better extrapolation, because there are basic differences in the skeletal elements of bird and pterosaur wings. In birds,

Figure 3.3. The giant pterosaurs from the Cretaceous period had wing-spans of up to 7 m. However, the proportions between the lengths of the various wing bones cannot be used for valid extrapolations from measurements on a single bone to the total wingspan for other flying reptiles. In general, allometric equations or regression lines apply only within the range of observations. From Gregory (1951) with courtesy of the Library Services Department, American Museum of Natural History.

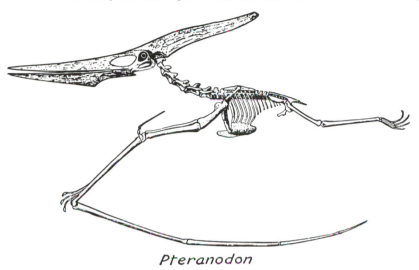

Pteranodon

the humerus is a relatively long bone, whereas the pterosaur wing has a different articulation in which the humerus is a short, squat bone. Both calculations depend on extrapolations, and allometric scaling should not be used for extrapolations, because the discontinuities and constraints are by definition unknown when we go beyond the range of known information.

Scaling fish

In the preceding discussion of brain size, we used body mass as the scale against which to measure brain size. Earlier we related metabolic rates to body mass. Even the scaling of the wingspan of the pterosaur has implications for the use of body mass (weight), because the wings must support the body (whatever their length may have been). Do we always refer our scaling problems to body mass?

Often this is the case, but other scales can be used and can lead to results that otherwise are not obvious. One case in which function is better understood if scaled against a linear dimension, rather than mass, concerns the swimming speed for fish. Speed is undoubtedly related to

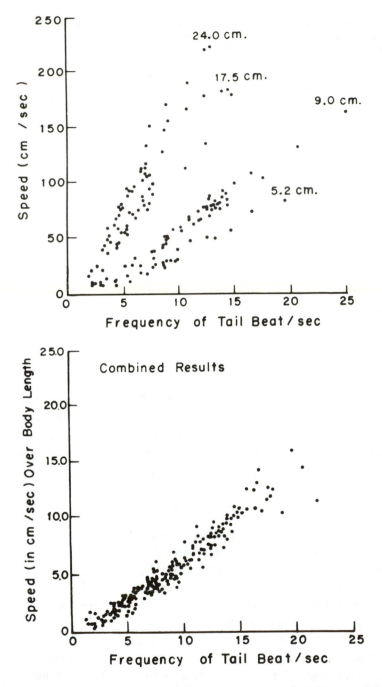

Figure 3.4. The swimming speed of a small freshwater fish, the dace, increases linearly with the frequency of the tailbeat. At any given

the body size of the fish; large fish swim faster than small fish, but the comparison looks different depending on what scale we use.

Swimming speeds for several kinds of fish were measured by Bainbridge (1958). One result was that the swimming speed, as expected, increased with increasing tailbeat frequency (Figure 3.4). The fact that the increase was linear was an interesting and less predictable finding. The linear relationship existed both for small and for large fish, but for any given tailbeat frequency, the large fish swam faster than the smaller.

If the speed, instead of being expressed in centimeters per second, is related to the body length of the fish, the resulting graph reveals a fundamental similarity between the small and the large fish that was not evident before (Figure 3.4). The ordinate of this graph expresses the fraction of its body length that a fish moves in 1 sec, and it has the dimension of time to the power −1, the same dimension as on the abscissa. This means that the slope of the regression line in Figure 3.4 is a dimensionless number that simply expresses that, regardless of the size of the fish, the distance moved for one beat of the tail is always the same fraction of its body length.

Once the analysis has been carried out, with the proper choice of scale to measure the size of the fish (in this case its length), the analysis reveals a simple and fundamental relationship that, in fact, is independent of the size of the fish. This does not apply to dace only, for other fish, such as trout and goldfish, show similar relationships between speed and tailbeat frequency. The slopes of the regression lines, however, differ from species to species, because of differences in swimming mechanics.

It would be possible to perform a similar analysis using the mass of the fish as a measure of its size. The mass of a fish is directly related to the third power of its linear dimensions, because small and large fish of the same species are virtually isometric. However, in the process of scaling swimming speed in relation to body mass, the more complex relationship will obscure the simplicity of the relationship just described. Also, we would not arrive at a simple dimensionless number as we did when using body length for scaling.

At times it may be difficult to see what scale is most useful. In the case of scaling fish, the proper choice revealed a simple relationship that is

Caption to Figure 3.4 *(cont.)*
frequency, a large fish swims faster than a small fish. However, if the swimming speed is calculated relative to the body length of the fish, the resulting plot reveals that the distance traveled for one beat of the tail is the same fraction of the body length for all fish, irrespective of size (Bainbridge, 1958). From Schmidt–Nielsen (1975a).

easily understood, and another choice of scale might have been less informative. In the rest of this book, nearly all variables will be scaled relative to body mass as the most suitable measure.

Although allometric equations express convenient and valuable generalizations, there are important limits regarding where they can and cannot be used, and the following points should be remembered:

1 Allometric equations are descriptive; they are not biological laws.

2 Allometric equations are useful for showing how a variable quantity is related to body size, all other things being equal (which most certainly they are not).

3 Allometric equations are valuable tools because they may reveal principles and connections that otherwise remain obscure.

4 Allometric equations are useful as a basis for comparisons and can reveal deviations from a general pattern. Such deviations may be due to "noise" or may reveal a significant secondary signal.

5 Allometric equations are useful for estimating an expected magnitude for some variable, an organ or a function, for a given body size.

6 Allometric equations cannot be used for extrapolations beyond the range of the data on which they are based.

4

How to scale eggs

A bird egg is a mechanical structure strong enough to hold the chick securely during development, yet weak enough to break out of. The shell must let oxygen in and carbon dioxide out, yet be sufficiently impermeable to water to keep the contents from drying out.

Bird eggs

Eggs are interesting structures. They are beautifully designed, self-contained life-support systems for the developing bird. All the nutrients, minerals, and water needed during incubation, as well as the necessary energy supply, are present in the freshly laid egg. This well-designed microcosmos contains everything needed for the growth and production of the hatchling chick, with one crucial exception: Oxygen. Furthermore, the shell of the avian egg is a simple physical system that is exceptionally well suited to considerations of scaling.

A hummingbird egg may weigh less than 0.3 g, and an ostrich egg over 1 kg, a 3000-fold range. The birds that lay these eggs range in size from 3-g hummingbirds to 100-kg ostriches, a 30 000-fold range. The largest bird that has ever lived, the elephant bird (*Aepyornis*) from Madagascar, was a sizable animal, standing perhaps 3 m tall and weighing over 500 kg (Feduccia, 1980). Its giant egg weighed about 10 kg, 10 times as much as an ostrich egg and 30 000 times as much as a hummingbird egg.

We are all familiar with hen's eggs. A newly laid egg may weigh 60 g, and if it is kept warm, occasionally turned, and kept at suitable humidity, it will be ready to hatch after 21 days, when its weight has decreased to 51 g. During that time span, the egg will have taken up 6 liters of oxygen (Figure 4.1) and given off 4.5 liters of carbon dioxide. It also will have lost water vapor by evaporation, reducing its total weight by about 9 g,

Figure 4.1. A newly laid chicken egg contains everything needed to form a complete hatchling chick, with the exception of oxygen. During the 21 days of incubation, the egg takes up 6 liters of oxygen, gives off 4.5 liters of carbon dioxide, and loses 11 liters of water vapor. The freshly laid egg weighs 60 g, the incubated egg when ready to hatch weighs 51 g, and the newly hatched chick weighs 39 g.

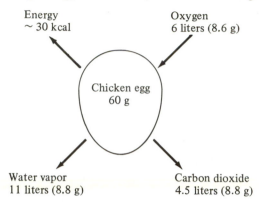

or 15% of its initial weight. The newly hatched chick will weigh 39 g, the remainder being eggshell and membranes, and the total energy consumption for making a chick from the fertilized egg will be about 30 kcal. This is nearly a third of the original energy content of the egg, which was about 100 kcal.

The chicken egg is a familiar example. What about eggs that are much smaller or much larger? We can inquire into the questions of size and scale from two viewpoints: either the size of the bird that laid the egg or the size of the egg itself. Both give interesting information.

Requirements to be met

What are the demands on the egg? First of all, it must have the mechanical strength to contain its liquid contents and support the weight of the parent during incubation, and it must tolerate being moved about in the nest without breaking. On the other hand, it must be thin enough to make it possible for the chick to break out of the egg at the end of the incubation period, because the parent bird does not help.

Next, the shell must permit passage of oxygen inward and carbon dioxide out. The material of the hard shell, calcium carbonate, is impermeable to gases, and therefore the gas exchange must take place through pores in the shell. The hard shell of a typical hen's egg is perforated by about 10 000 pores, each with a functional diameter of about 17 μm.

However, the evaporation of water from the egg must be kept within reasonable bounds, carefully adjusted to the incubation time. These contrasting demands evidently have been successfully met, and the appropriate variables (pore size and numbers) are scaled to correspond to the demands of eggs that differ in mass by a factor of some 30 000.

Many of the questions pertaining to the scaling of bird eggs have been answered by a long series of careful studies and extensive reviews of the large literature on eggs, carried out by Professor Hermann Rahn and his collaborators at the University of Buffalo. Much of what I have to say is based on Rahn's results, and I shall use few references to the earlier literature, which is both extensive and valuable.

Egg size and bird size

Data on the size of the eggs of 800 species of birds was reviewed by Rahn and associates (1975), based on information accumulated by some of the world's most distinguished ornithologists. They used the method that was suggested by Huxley (1927), which was to plot egg weight against body weight on logarithmic coordinates. For many different orders and families, the regression lines tended to have similar slopes, averaging 0.675. However, the proportionality factor (intercept at unity body mass) was characteristically different for each group.

This last situation raises a difficult question that cannot readily be answered on theoretical grounds: Is it correct to accept the characteristic slope of about 0.67 for the different groups as representing birds in general, or should one combine all eggs that have been measured in one big pool and calculate the overall slope for all bird eggs? If this is done, the common regression equation for egg mass (M_{egg}) on body mass (M_b) in grams is

$$M_{egg} = 0.277 \, M_b^{0.770} \tag{1}$$

Why the different slopes? The discrepancy is easily explained. Each regression line for a different group of birds covers a different size range, and although the lines are parallel, smaller birds tend to have a lower intercept (perhaps because of larger clutch sizes), whereas larger birds tend to have a higher intercept. This will tend to rotate the overall line in a counterclockwise direction, thus increasing its slope.

Rahn and his collaborators sought a reasonable answer in the following way. They tested the slopes of the individual regression lines for non-parallelism by analysis of variance, which showed that the individual slopes did not differ significantly from one another ($p > 0.99$). The common slope had a value of 0.675, with 95% confidence limits of ±0.015.

It may seem reasonable to accept this common slope, $b = 0.675$, which is valid within each order of birds, as well as valid for all orders of birds. However, that brings us into the difficult situation that if we consider all eggs, the results will be misleading in the lowest and highest ranges of bird size. This is a dilemma that has no "right" answer.

Incubation time

The time needed for incubation, the incubation period, is much less variable than the size of eggs. The incubation period may be as short as 11 days or as long as nearly 90 days, a factor of about 8. By plotting the incubation period, or incubation time (t_{inc}, in days), as a function of body mass of the parent bird (M_b, in g), Rahn and associates (1975) obtained the following equation:

$$t_{inc} = 9.105 \, M_b^{0.167} \tag{2}$$

The slope is numerically small, in accordance with the moderate effect of bird size on incubation time.

The two variables we have considered, egg size and incubation time, are plotted together in Figure 4.2, but in this case the egg weight is given as percentage of body weight. One aspect of this graph must be noted: The ordinate scale is arithmetic rather than logarithmic, although the abscissa gives the body mass on the customary logarithmic scale. The body size range is from 2.5 g, the lower size limit for hummingbirds, to about 1000 kg, the size of the largest elephant birds.

Once the egg is laid and is being properly incubated, it is an independent living system, totally unconcerned with the size of the bird it came from. We have surmised that, as a rule, smaller eggs hatch sooner than large eggs, a relationship that is shown in Figure 4.3.

Rahn and Ar (1974) used data from 475 species of birds and obtained the following equation for incubation time (t_{inc}, in days) relative to egg size (M_{egg}, in g):

$$t_{inc} = 12.03 \, M_{egg}^{0.217} \tag{3}$$

This equation could, of course, be readily calculated from the two preceding equations, which related egg size and incubation time to the body size of the parent bird. By relating these two variables, incubation time and egg size, directly to each other, the arithmetic result is identical with that of the preceding equation.

Figure 4.2. Larger birds lay proportionately smaller eggs, and the incubation time for an egg increases with increasing body size of the parent. When the egg size is expressed as a percentage of the adult bird weight, egg weight becomes an increasingly smaller fraction of body mass for larger birds. The dotted lines span the 68% confidence limits, which indicate that the variability is substantial. Note that the ordinates for egg weights and incubation days both are on arithmetic scales. From Rahn et al. (1975).

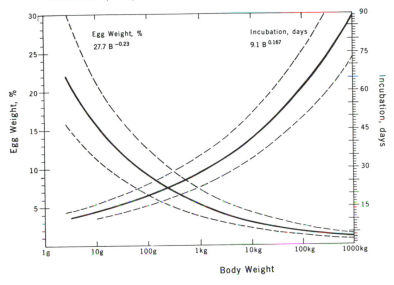

Pores in the eggshell

Let us consider the pores that permit passage of the respiratory gases as well as water vapor through the hard shell. The typical chicken egg is perforated by about 10 000 pores. The surface area of the egg is about 70 cm², which gives an average of about 1.5 pores per square millimeter of shell. The pore diameter is about 17 μm, so that the total pore area of a hen's egg is 2.3 mm² (Wangensteen et al., 1971). The thickness of the shell of a hen's egg is about 0.35 mm. It is now a fairly simple matter to calculate the diffusivity of gases through these pores; the result is that the pores suffice to supply the egg with oxygen by diffusion through the pores at the rate required just before hatching, when the rate of oxygen consumption is highest.

How do pore length and diameter vary with the size of the egg? Obviously, larger eggs have thicker shells, and shell thickness determines the length of the pores. The thickness of the shell (pore length, L_{pore},

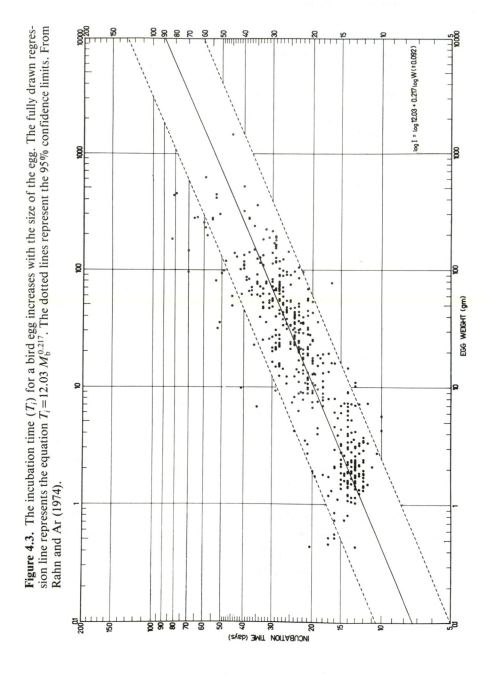

Figure 4.3. The incubation time (T_i) for a bird egg increases with the size of the egg. The fully drawn regression line represents the equation $T_i = 12.03\,M_b^{0.217}$. The dotted lines represent the 95% confidence limits. From Rahn and Ar (1974).

in mm) varies with egg size (M_{egg}, in g) for 367 species of birds according to the following equation (Ar et al., 1974):

$$L_{pore} = 5.126 \cdot 10^{-2} M_{egg}^{0.456} \qquad (4)$$

When it comes to pore diameter we are in serious trouble, because the pores are not neat, cylindrical structures; they are highly irregular. The way around this difficulty is to determine the total pore area experimentally from the diffusion of gases through the eggshell. If we know the diffusion coefficient, it is possible to calculate a functional pore area, which is precisely what is of interest for the gas exchange. It is therefore not necessary to examine the precise geometry of the pores and their numbers.

It is obvious that the total pore area must increase with the size of the egg, and the increasing pore length (with increased shell thickness) adds to the requirement for an increase in pore area. By using the rate of diffusion of water vapor through the eggshell, Ar et al. (1974) calculated the functional pore area (A_{pore}, in mm^2) to be

$$A_{pore} = 9.2 \cdot 10^{-3} M_{egg}^{1.236} \qquad (5)$$

What do the equations that relate pore length and area to egg size mean in terms of real eggs? Consider an ordinary chicken egg of 60 g and another egg 10 times as large (which could be the 600-g egg of the rhea, a South American ostrichlike bird). The 10-fold larger egg will have a total functional pore area that is about 17 times larger ($10^{1.236} = 17.2$). This, of course, increases the gas conductance, which most certainly is needed for the larger developing chick in the larger egg. However, the larger egg has a thicker shell, and the pore length is therefore increased, being about 2.9-fold longer in the larger egg ($10^{0.456} = 2.86$).

The gas conductance is proportional to the functional pore area and inversely proportional to the pore length (diffusion distance), and the gas conductance will be six times larger in the egg of the rhea than in the chicken egg one-tenth its size. This corresponds to our expectation that the chick in a 10 times larger egg should have a sixfold higher metabolic rate (see Chapter 6).

Water loss from eggs

Because the ease with which gases, including water vapor, diffuse through the pores is directly related to the pore area and inversely related to the pore length, the conductance for gases should be related to the ratio between these two variables. Using the foregoing equations, we find that gas conductance (G) should be proportional to egg size as follows:

$$G \propto M_{egg}^{1.236} : M_{egg}^{0.456} = M_{egg}^{0.780} \tag{6}$$

This is indeed the way the observed water vapor conductance, and thus the rate of water loss from the egg, is related to egg size (Ar et al., 1974).

The equation for water vapor conductance,[1] based on measured diffusion of water vapor through eggshells from 29 species, with egg weights ranging over three orders of magnitude, is

$$G_{H_2O} = 0.432 \cdot M_{egg}^{0.780} \tag{7}$$

Because the water vapor conductance of the whole egg increases with egg mass raised to a power less than unity, this means that the rate of water loss per gram of egg decreases as the egg gets larger; in other words, large eggs lose relatively less water than small eggs. However, as we saw before, birds with large eggs tend to have longer incubation periods, and the eggs are therefore losing water over a longer period of time.

The combined effect of the lower relative evaporation and the longer incubation period is easy to calculate and gives a strikingly simple result. Total water loss during the entire incubation is the product of the daily water loss (directly related to water vapor conductance) and the incubation time. The former is related to egg weight to the power 0.78, and incubation time to egg weight to the power 0.22; that is, total water loss is related to egg weight to the power 1.00. This says that all eggs, irrespective of size, lose water in direct proportion to the size of the egg. In other words, the fraction of egg weight that is lost through evaporation during the incubation is the same fraction for all eggs, irrespective of egg size. Under normal conditions, with a water vapor pressure in the nest of about 35 mm Hg, all eggs should normally lose about 15% of their initial weight through evaporation during the normal incubation period.

To summarize, we have seen that pore length (thickness of the eggshell) and the pore area available for diffusion are adjusted over the entire 30 000-fold difference in egg size, so that the water loss of all eggs during a normal incubation is the same fraction of the initial egg weight, while the respiratory gases can diffuse through the pores at the rate required by the developing embryo.

---

1 The water vapor conductance of the eggshell is defined as $G_{H_2O} = \dot{E}_{H_2O}/\Delta P_{H_2O}$, where $\dot{E}_{H_2O}$ is the rate of evaporation in milligrams per day and ΔP_{H_2O} is the water vapor pressure difference across the shell.

We have now considered the size of bird eggs, the incubation time, and the structure of the shell as related to gas exchange and water loss. An essential function of the eggshell, to serve as a mechanical protection for its liquid contents and the developing embryo, will be discussed in the next chapter.

5

The strength of bones and skeletons

What skeletons do

Skeletons are support systems; they keep animals from collapsing, and they provide levers on which muscles can act. Skeletons are hard and mechanically rigid, and they can be either internal structures (endoskeletons, as in vertebrates, from fish to mammals) or external (exoskeletons, as in arthropods, from crabs to spiders and insects).[1]

Skeletons are mechanical structures that must withstand the forces that impinge on them, or they fail. The skeleton must support the weight of the animal, or it will fail in compression by crushing. It must withstand the forces due to locomotion, which produce bending and torsion that may cause failure in buckling. Also, a skeleton must withstand the forces of impact, and this may be the most critical requirement.

A heavier skeleton will be stronger, and for animals that do not move about (corals, for example), most of the organism can consist of skeletal material. For animals that move about, a heavy skeleton increases the cost of moving about and also reduces agility and the chances of escape from predators. The size and structure of the skeleton will therefore involve a compromise between the various demands: What should be optimized, strength or lightness?

1 A third kind, hydrostatic skeletons, is based on a different principle. In these, the forces act on a compression-resisting element, which is a fluid confined in a tension-resisting container. Many invertebrates (e.g., worms) derive their shape and their ability to move from the combined effects of the body fluids and the muscle layers in the body wall. All of us have seen the principle of a hydrostatic skeleton in the wilting of leaves and flowers. Normally these are kept erect by the cell content exerting a pressure against the unyielding cell walls; if the fluid volume is reduced through water loss, the pressure is reduced and the plant droops.

Figure 5.1. Galileo was probably the first to point out that the bones of very large animals must be scaled out of proportion to their linear dimensions in order to support the weight of the animal, which increases with the third power of the linear dimension. From Galilei (1637).

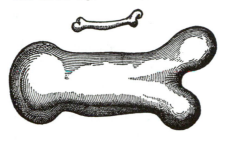

Scaling mammalian skeletons

Consider a group of vertebrates, mammals, for example, that are similarly organized, although they are not alike. It is familiar to all of us that the bones of an elephant are proportionately heavier and stockier than those of a mouse. Why?

The legs of the typical mammal support its body weight, and as the weight of the animal increases, the strength of the support must be correspondingly increased. Assume that all linear dimensions of an animal are increased by a factor of 2. The mass of the similarly shaped larger animal is then increased by a factor of 8, the cube of the linear dimension. This increase in weight must be reflected in the strength of the supporting structures. To avoid crushing, their cross-sectional area must be increased in proportion to the eightfold increase in load, but if all dimensions are merely doubled, the cross-sectional area of the bones will only be quadrupled. This will not suffice, and to support the eightfold increase in weight, the bones must be scaled out of proportion.

This need for a disproportionate increase in the size of supporting bones with increasing body size was understood by Galileo Galilei (1637), who probably was the first scientist to publish a discussion of the effects of body size on the size of the skeleton. In his *Dialogues* he mentioned that the skeleton of a large animal must be strong enough to support the weight of the animal as it increases with the third power of the linear dimensions. Galileo used a drawing (Figure 5.1) to show how a large bone is disproportionately thicker than a small bone. [Incidentally, judging from the drawing, Galileo made an arithmetical mistake. The larger bone, which is three times as long as the shorter, shows a 9-fold increase in diameter, which is a greater distortion than required. A 3-fold increase in linear dimensions should give a 27-fold increase in mass, and the cross-sectional area of the bone should be increased 27-fold, and its diameter therefore by the square root of 27 (i.e., 5.2 instead of 9)].

If the mass of the skeleton increases out of proportion to the increase in the body mass, there must be a limit beyond which a further increase is impossible, for the whole animal cannot be skeleton. This was understood by Galileo, who suggested that giant-sized animals must be aquatic so that their weight can be supported by water. On land, their skeletons would collapse under their enormous weight, and the largest animals, the whales, are therefore aquatic. In Galileo's presentation, Simplicio raised the question of ''the enormous size reached by certain fish, such as the whale, which, I understand, is ten times as large as an elephant.'' In the answer, Salviati pointed out that in spite of the enormous weight of their bones, these animals do not sink, and ''The fact then that fish are able to remain motionless under water is a conclusive reason for thinking that the material of their bodies has the same specific gravity as that of water; accordingly, if in their make-up there are certain parts which are heavier than water there must be others which are lighter for otherwise they would not produce equilibrium'' (Galileo, 1637).

Consider the four legs of a mammal as vertical columns that must support the weight of the animal. The first question is whether the mechanical properties of bone are the same in large and small animals. This is probably so, because bones of all mammals are made of the same material, crystalline calcium apatite embedded in a matrix of collagen. Measurements of failure stress in bones from mammals ranging from 0.05 to 700 kg (a 14 000-fold range) showed no significant differences (233 ± 53 MN/m^2 for small animals and 200 ± 28 for large animals) (Biewener, 1982).

Because the compressive strength of bone in larger animals cannot be increased above the limit set by the material, increased strength must be achieved by changing the dimensions of the bones. To increase the strength of the supporting columns in proportion to the load, their cross-sectional area must increase in proportion to the load (M_b). Next, let the length of the supporting column increase in proportion to a characteristic linear dimension (i.e., in proportion to $M_b^{0.33}$). The volume, or mass, of the supporting column will then be the product of its cross-sectional area ($M_b^{1.0}$) and its length ($M_b^{0.33}$), that is, proportional to $M_b^{1.33}$.

It is a simple matter to express these numbers in ordinary language. To support the weight of an increasing mass of a given shape, a column must increase out of proportion to the linear dimensions of the load. If the height of the column is scaled in proportion to the linear dimensions of

Figure 5.2. The mass of the mammalian skeleton increases out of proportion to an increase in body mass, as would be expected for theoretical reasons. However, the slope of the empirical regression line, 1.09, is less than expected from considerations of the compressive strength needed to support the weight of the body. From Prange et al. (1979).

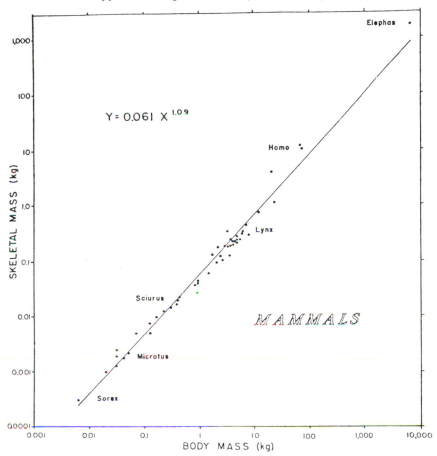

the load, the volume or mass of the supporting column should be scaled as the load raised to the power 1.33. This is not the way the skeletons of real animals are scaled.

What about real animals?

A great deal of information about mammalian skeletons was compiled by Prange and associates (1979), based on a variety of sources in addition to their own observations. The resulting graph is shown in Figure 5.2. These authors published the original data in table form, and we can therefore calculate not only the slope but also the confidence limits. The slope is 1.08, with 95% confidence limits of ±0.04; that is, the slope is between 1.04 and 1.12. This is significantly above the slope of 1.00 that describes proportionality, but it is far from the "expected" slope of 1.33 derived from the consideration of static loads.

What does this mean? If the skeleton of a small animal is designed to support the gravitational load and the slope is only 1.08, the elephant must be greatly underdimensioned. This is not likely. On the other hand, if the pillarlike legs of the elephant provide adequate support for its weight, then small mammals would appear to be greatly overdimensioned for the gravitational loads they must bear.

The conclusion invites itself: The skeletal mass of a mammal is not scaled primarily to support gravitational loads, and we must look elsewhere. During locomotion, there are always forces due to rapid acceleration and deceleration that expose the skeleton to stresses of bending and torsion, and the skeletal elements must be able to resist failure by primary buckling. Because the shapes of the skeletal elements are highly complex, the added stresses on the skeleton during acceleration and impact loading become extremely complex (Currey, 1967).

Some estimates of the stresses on the leg bones of mammals during locomotion were made by Alexander and associates (1979b). The maximum stresses in the bones of the legs were calculated for hopping kangaroos, jumping dogs, galloping antelope and buffalo, and running elephants. The body masses of these animals ranged from 7 to 2500 kg, a 350-fold range; yet the estimated bone stresses were amazingly similar, around 50 to 150 MN/m^2 (500–1500 kgf/cm^2), without any clear relationship to body size. The range was extended to 0.1 kg body size by Biewener (1983), who calculated stresses of 58 to 86 MN/m^2 in the distal limb bones of chipmunks and ground squirrels.

Considering the difficulty in calculating bone stresses during locomotion, we must conclude that there are no discernible differences in the maximum bone stresses in animals over a 25 000-fold size range. In other words, the maximum stresses that occur during the most vigorous forms of locomotion approach the yield strength of the bones, which is 170 MN/m^2 in tension and 280 MN/m^2 in compression (Burstein et al., 1972). This, of course, is consistent with the common occurrences of sprains and fractures of bones in human athletes during maximum performance; extreme athletic training apparently brings the human organism close to and at times beyond its design limits.

It is interesting that the maximum tensile stresses estimated by Alexander were consistently somewhat lower than the compressive stresses by about one-third or so. Is it merely a coincidence that the tensile and compressive strength of bone (170 and 280 MN/m^2, respectively) differ in roughly the same ratio as the maximum tensile and compressive stresses reported by Alexander?

How light are bird bones?

Most of us know that the bones of birds are lighter than mammalian bones. Right? Those who are better informed may also know that many bird bones, especially the wing bones, are hollow and filled with air. It is therefore a surprise to find that the total mass of bird skeletons relative to body mass does not differ much from mammalian skeletons. The equations for skeletal mass (M_{skel}, in kg) relative to body mass (M_b, in kg), based on the data of Prange et al. (1979), are

$$\text{birds:} \qquad M_{skel} = 0.0649\, M_b^{1.068 \pm 0.008\ \text{SE}}$$

$$\text{mammals:} \qquad M_{skel} = 0.0608\, M_b^{1.083 \pm 0.021\ \text{SE}}$$

These two equations are virtually identical. We can immediately see that typical birds and mammals of 1 kg body mass will have skeletons of 65 and 61 g respectively. Thus, birds may have very slightly heavier skeletons than mammals of the same body size, although the difference is not statistically significant.

This similarity does not preclude the possibility that certain individual bones may be relatively lighter in birds, and this holds true for the long wing bones, which are indeed lighter than the arm bones of mammals. The leg bones, in contrast, are relatively heavier in birds than in mammals. Is this because the forces of impact during landing are absorbed by two rather than four legs?

The information that was compiled by Prange and associates referred to air-dried bones. The avian skeletons were from museum material, and in a later article, Prange and his collaborators (Anderson et al., 1979) stated that all the land-mammal skeletons were "typical air-dried preparations characteristic of museum collections," thus implying that they were comparable to the bird material.

The differences between fresh and dry bird skeletons were studied by Dosse (1937), who found that the dry weight of bird skeletons averaged 70.2% (± 5.5 SD, $N = 22$) of the fresh weight. There were no significant variations in the dry weight of the bones with the body size of the bird, and the equations given earlier for dry bones can therefore be recalculated for fresh weight by adding roughly 50% to the dry weight. This will bring the numerical factor from about 0.065 to nearly 0.1, a number that is in line with the mass of fresh bird skeletons reported by other authors.

Aquatic animals: lighter skeletons?

Aquatic animals have their weight supported by water. Whales and seals are nearly neutrally buoyant in water, and it is reasonable to ask if this is reflected in their skeletons. In these animals the skeleton serves to provide the stiffness needed for muscle action, but not to support the body weight.

The two most prominent groups of aquatic mammals are seals and whales, which together range over more than four orders of magnitude in size. Unfortunately, very little information can be found about the size of seal skeletons, but such information would not be too important for our question, because seals are not fully aquatic, spending considerable periods of time on land.

Information about whales, on the other hand, is readily available because of commercial interest in whaling, and a great deal of information has been compiled by Smith and Pace (1971). On the basis of 170 specimens of whales of six different species, the following equation was calculated for the mass of the skeleton (M_{skel}) in relation to body mass (M_b), measured in metric tons:

$$M_{skel} = 0.105 \, M_b^{1.107}$$

The exponent 1.107 is certainly similar to the exponent for the skeleton of land mammals, 1.08. The proportionality coefficient, 0.105, is higher than for land mammals (0.0608). However, the whale bones undoubtedly were weighed fresh, whereas the equation for land mammals referred to dry bones; in addition, the mass unit is tons in the whale equation and kilograms for land mammals.

If the whale equation is recalculated to the same units as used for land mammals, kilograms for both skeletal and body mass, the proportionality coefficient will be 0.050. This will give a 1-kg whale a skeleton of 5% of the body mass, which is meaningless, because no whale is that small. However, it confirms that regression equations apply only within the range of observations and are not valid for extrapolation.

There is another difficulty in evaluating the preceding equation. It was based on all the available specimens, but when only mature whales are considered, the equation looks as follows:

$$M_{skel} = 0.137 \, M_b^{1.024}$$

The proportionality coefficient is surprisingly high, but the striking difference is that the exponent is now 1.024, not significantly different from 1.00. In other words, based on observations of only mature specimens, the skeleton of whales may scale strictly in proportion to body size. This suggests that when there is no need to support the body in the gravitational field, the skeleton is not required to scale out of proportion to body size.

Unfortunately, no information seems to be available for the smallest whales: dolphins and porpoises. Such information would extend the size range by two orders of magnitude and would make it possible to arrive at a more firm conclusion about the effect of gravity on the skeleton.

Perhaps we can look elsewhere for additional information. Fish are also vertebrates with endoskeletons and have their weight fully supported by water. They range in size from milligrams to tons and could provide very interesting information. However, I have found only one study of fish skeletons (Reynolds and Karlotski, 1977). The dry weight of the skeleton was determined for 11 fish (five species) ranging from 3 to 1200 g. The regression equation, expressed in gram units, was

$$M_{skel} = 0.033 \, M_b^{1.03}$$

The exponent in this equation, 1.03, is not significantly different from 1.00 (SE = 0.03), which tends to confirm that the absence of gravitational effects is responsible for the scaling.

The numerical factor for the skeletal mass of fish, 0.033, is lower than for mammals. The reason is, in part, that the bones were prepared by boiling, soaking in ammonia, and oven drying, and therefore they had lost both oils and water. Even so, it seems that the fish skeleton, as compared with those of birds and mammals, is only about half as heavy. Consider, however, that fish have an entirely different body shape and form of locomotion, which makes the comparison of doubtful value.

To summarize the modest information we have about aquatic vertebrates, we can say that in mature large whales and in a small group of teleost fish, the skeleton apparently scales in proportion to body mass. If this should be confirmed when additional material becomes available, it can best be interpreted as the result of different gravitational effects on land and aquatic animals. However, for the time being, the question must be considered unresolved.

The strength of bones

Instead of examining the entire skeleton, in which hundreds of elements serve a wide variety of complex functions, consider only the leg bones of four-footed mammals. These can be regarded as rods or levers that must be able to resist failure under gravitational loads and especially loads due to acceleration, deceleration, and impact.

Mechanical failure in a loaded column may occur in either of two ways. The column may yield in compression, crushed by the weight placed on it, or it may fail in primary buckling, as when a heavy load is placed on top of a tall slender column. If the stiffness of such a column is insufficient to resist bending, it will buckle and fail.

When we speak of stiffness and bending, we are actually discussing the mechanical property known as elasticity. This is the property that allows a body to recover its shape following a distortion caused by a stress or a load imposed on it. Failure in primary buckling is therefore known as failure in elastic buckling.

Now consider the scaling of the supporting columns as the load (the size of the animal) is increased. The columns (the bones of the limbs) could be scaled in either of three ways: in *geometric similarity,* in *static stress similarity,* or in *elastic similarity.*

We discussed earlier that scaling in geometric similarity (isometric scaling) will lead to failure because the compressive strength of the supporting column increases only with the square of the diameter, while the load increases with the third power of the diameter. The discussion of scaling according to similarity in static stress led to the conclusion that the cross-sectional area of the supporting column should be scaled in proportion to the load (body mass) and its diameter as the body mass to the power 0.5. The conclusion was that the supporting bones should be scaled as body mass to the power 1.33. Because the mass of the skeleton does not scale according to this alternative, it was concluded that static stresses are not the primary factor in the scaling of mammalian skeletons.

The third possibility, scaling according to elastic similarity, turns out to be an extremely interesting hypothesis. The theory, which considers a tall, slender cylindrical column, was developed from basic principles by McMahon (1973). If the column is slender enough, it may fail in elastic buckling; that is, a small lateral displacement allows its weight to exert a lateral moment of force that causes the column to bend further, thus increasing the moment of force and causing the column to fail. The critical variables (in addition to the elastic modulus of the material) will be the length and the diameter of the column.

McMahon analyzed the scaling of trees by considering them as tall tapering columns. By examining the dimensions of trees and their natural frequencies of oscillation, he found that trees of several different species preserve elastic similarity as they grow in size (McMahon, 1975a; McMahon and Kronauer, 1976).

Extending the argument to animals, in which the body mass presumably scales with the third power of the linear dimensions, McMahon showed that in any supporting bone, the diameter (d) should scale with the $\frac{3}{8}$ power of the body mass (M_b), and the length (l) should scale with one-quarter of the body mass, or

$$l \propto M_b^{1/4} \quad \text{and} \quad d \propto M_b^{3/8}$$

McMahon then examined the limb bones of adult ungulates over a wide size range (McMahon, 1975b). He measured length and diameter of major limb bones in 118 museum skeletons. The lengths of the bones differed by a factor of 7; if geometric scaling prevails, the body masses should differ by a factor of 7^3, or 343; according to elastic similarity, the body masses should differ by a factor of 7^4, or 2400. As is common for museum specimens, body weights had not been recorded, but the proportions of the bones were indeed scaled in close agreement with McMahon's hypothesis. His model predicts that bone length, l, should scale with bone diameter to the power 0.67. The measured bones were in substantial agreement with this model, because the exponent was close to 0.67 for the forelimbs and slightly less for the hindlimbs.

McMahon's study concerned a very homogeneous group of mammals, the ungulates. A similar study by Alexander and associates (1979a) included the entire range of land mammals, from shrew (2.9 g) to elephant (2500 kg). Alexander used fresh material, but this does not affect the measured linear dimensions of the bones. If McMahon's theory of elastic similarity is correct, the length of the limb bones should scale with the body mass to the power 0.25. Alexander found that the lengths of corresponding limb bones from 37 species of mammals tended to be proportional to the body mass to the power 0.35, and their diameters to the body mass to the power 0.36, except in the family Bovidae (ungulates), in which the exponent for the length was closer to the value of 0.25 predicted by McMahon's theory of elastic similarity. Thus, McMahon's measurements were confirmed for the animal group he had examined, but his theory was not supported for material extending over the entire size range of mammals.

The overall results of Alexander's study (1979a) are shown in Table 5.1. Because the volume of a bone is proportional to the product of its length and its diameter squared, we see that the volume of the bones will be proportional to body mass to the exponent 1.07:

Table 5.1. Lengths (l) and diameters (d) of mammalian leg bones related to body mass (M_b) through the equations $l = a M_b^b$ and $d = a' M_b^b$; Alexander et al. (1979a).

	Mean exponent (b) for:	
	Length	Diameter
Insectivora	0.38[a]	0.39
Primates	0.34	0.39
Rodentia	0.33	0.40
Fissipedia	0.36	0.40
Bovidae	0.26	0.36
All mammals	0.35	0.36

[a] The exponents are means based on all the measured bones (femur, tibia, metatarsal, humerus, ulna, and metacarpal) for each group.

$$\text{volume} \propto l \cdot d^2 = M_b^{0.35} \cdot M_b^{0.36 \cdot 2} = M_b^{1.07}$$

The exponent 1.07 for the bone volume is identical with the exponent for skeletal mass in mammals that was discussed earlier, and it suggests that similar principles must be involved in the scaling of the leg bones and the entire skeleton.

An interesting question remains: Why do the bovids seem to scale according to the theory of elastic similarity, whereas mammals in general do not? Perhaps it is because the bovids stand on fairly straight legs, and the front legs, in particular, are pillarlike. Ungulates are quite similar from the smallest to the largest, and the shape of their legs conforms to the column structure that was the basis for McMahon's theory. Among the other groups, insectivores, primates, rodents, carnivores, and so forth, the usual posture does not provide straight, vertical legs. Perhaps this explains the difference in scaling. The theory of elastic similarity may very well apply to the structure of ungulates but not to mammals in general.

External skeletons: a complicated matter

Animals with external skeletons (crustaceans, insects, and other arthropods) cover a tremendous size range. The smallest insects weigh less than 25 μg, and the largest weigh more than a million times that figure.

However, insects never attain the large size of the largest crustaceans, which are aquatic. We may recall that the legs of the Japanese spider crab may span as much as 4 m. Obviously, no terrestrial arthropod comes anywhere near this size, although some land crabs are quite large.

The exoskeleton of arthropods serves several functions. It gives animals their external shape, and in air (as opposed to water) the larger arthropods need rigid skeletons to retain their shape. Next, the skeleton provides support for the muscles, as it does for animals with an endoskeleton. Finally, the exoskeleton provides protection against forces of impact and against predators. As a result, the exoskeleton must resist not only bending and buckling forces but also local stresses of impact. This makes the analysis of its mechanical properties extremely complex.

The relative merits of internal and external skeletons have been discussed by Currey (1967). A mechanical analysis, based on fundamental principles of mechanics, suggests that an exoskeleton is superior to an endoskeleton in regard to failure by buckling and bending. When it comes to external forces of impact, however, the situation is different; here the exoskeleton is at a disadvantage, particularly for large and active animals such as vertebrates. The kinetic energy that is absorbed on impact is proportional to the velocity squared ($E = \frac{1}{2}MU^2$), and this makes the velocity of impact particularly important. An endoskeleton is much less likely to fail through impact, because the soft tissues can absorb a great amount of energy without serious damage, whereas the hard, stiff exoskeleton is unprotected. The entire kinetic energy is absorbed on impact, and for a fast-moving large animal, the forces impacting on a hard exoskeleton are likely to cause local failure.

Spiders compose a rather uniform group of arthropods with similar geometrical shapes. The spider exoskeleton has been studied over a size range from 25 to 1200 mg (Anderson et al., 1979). The equation for the mass of the supportive tissue (exoskeleton) is the following (skeletal mass, M_{skel}, and body mass, M_b, in g):

$$M_{skel} = 0.078 M_b^{1.135}$$

A couple of comments will help clarify the meaning of this equation. First, for a 1-g spider, nearly 8% of the body mass will be in the skeleton, a proportion similar to that in the average-sized mammal. Second, the fraction that is in the skeleton increases with increasing body size (in fact, faster than for mammals). The exponent for spiders, 1.135, is somewhat higher than for mammalian skeletons (1.08), but the difference may not be significant.

Breaking eggshells

What about the "exoskeleton" of an organism that never moves about of its own accord, the avian egg? Because of its regular shape, uniform thickness, and constant composition, the shell of the egg should be much simpler to analyze. The eggshell consists primarily of calcium carbonate, and it has a density very nearly 2.0 g/cm^3. (There is a slight but significant increase in the density of the shell with size, from 1.95 for a 1-g egg to 2.14 for a 1-kg egg, but we can disregard this difference.) Information about the eggs of 368 species of birds was compiled by Paganelli et al. (1974), and the resulting equation for the mass of the shell (M_{shell}, in g) was

$$M_{shell} = 0.0482\, M_{egg}^{1.132}$$

This equation predicts that, on the average, the smallest egg, that of hummingbirds, which weighs about 0.3 g, should have a shell weight 4% of the fresh egg weight. The egg of the largest living bird, the ostrich, weighs about 1 kg and should have a shell weight of 120 g, 12% of the egg weight. Relative to the egg of the hummingbird, this is a threefold increase in the relative weight of the shell.

Because the surface area of a bird egg is an extremely close function of the square of the radius, or the volume to the 0.67 power, the entire change in the mass of the shell is due to a relative increase in the thickness of the shell.

The shell not only provides protection but also must permit the exchange of gases (oxygen and carbon dioxide), as discussed elsewhere in this book. Gas exchange takes place through pores, but the thickness of the shell probably is not a constraint on gas diffusion, because an increased pore length can readily be compensated for by slight changes in pore diameter. We can therefore consider that the thickness of the shell is entirely related to the mechanical properties of the bird egg and its resistance to external forces: gravity and impact. However, the chick must be able to break out of the shell at the end of incubation, and this is a specific constraint on increasing shell thickness.

What are the external forces that may break an egg? The eggs must support the weight of the parent bird, and as the eggs are moved around in the nest by the incubating bird, there is some impact. Probably impact is the more important consideration, because the forces are concentrated at the points of contact with other eggs, whereas the weight of the parent is supported over a larger area of the egg.

There is a striking similarity in the exponent for shell mass versus egg mass (1.132) and the exponent for the mass of the spider exoskeleton relative to body mass (1.135). This means that the thickness of the eggshell scales relative to its volume with an exponent that is identical with the exponent for the thickness of the spider exoskeleton relative to body mass (assuming that large and small specimens within each group are geometrically similar). Whether this is coincidence or an indication of a fundamental similarity in the mechanical requirements cannot be determined until the highly complex system of the arthropod exoskeleton is better understood.

An evaluation of the relative strength of eggs of various sizes is informative. The load required to initiate crushing of an egg was measured by mounting the egg vertically between two polished metal plates and applying increasing force to the egg (Ar et al., 1979). The crushing force (F, in gram force) as a function of egg size (gram mass) was the following:

$$F = 50.86 \, M_{egg}^{0.915}$$

We can see that the strength of the eggshell increases with the size of the egg, but not quite in proportion to its size, because the exponent is less than 1.0. One could therefore say that a larger egg, relative to its mass, is less strong than a smaller egg. Now consider that the cross-sectional area of the eggshell is a function of the square of the shell's thickness and that the thickness scales with the 0.46 power of the egg mass. The cross-sectional area of the eggshell therefore scales with the exponent 0.92. This is precisely the same exponent as for the crushing load; that is, the crushing load is a direct function of the square of the shell's thickness, as could be expected from a mechanical analysis of the problem (Ar et al., 1979).

The weight of the incubating bird, although it does not directly crush the egg, is of importance because of the forces it exerts at the points of contact with neighboring eggs as they are moved about in the nest. Because larger eggs are laid by parent birds that are disproportionately large relative to the egg, the large egg is pushed against its neighbors by a disproportionately heavy bird. However, failure of the normal eggshell in the nest is uncommon; so the weight of the parent bird may be unimportant. The constraint on acceptable shell thickness is probably the force the hatchling chick is able to exert in breaking out of the shell, not the external forces on the egg during incubation.

In this chapter we have been concerned with the scaling of structures. In the next chapters we shall turn to the scaling of functions, beginning with the relationship between metabolic rate and body size.

6

Metabolic rate and body size

In earlier chapters we were concerned with the scaling of structures; in this and later chapters we shall deal mostly with function. The first subject will be the metabolic rates of animals. Energy is needed for maintenance and for all the normal functions of the living animal: for moving about, for feeding, for escaping, and so on. The energy an animal needs for all this comes from the chemical energy contained in food. The total use or turnover of chemical energy is frequently referred to as the metabolic rate, and for reasons that we shall not discuss here, it is convenient as well as reasonably accurate to measure the rate of energy metabolism as the rate of oxygen consumption.

The determination of oxygen consumption is technically easy, and it is so commonly used for estimation of metabolic rate that the two terms often are used interchangeably. This is not correct; for example, an anaerobic organism that depends on nonoxidative metabolic processes has zero oxygen consumption, but it certainly does not have a zero metabolic rate. In the following, however, we shall be concerned mostly with the rate of oxidative metabolism as it almost universally is measured as the rate of oxygen consumption.

In the absence of external activity, metabolism, or oxygen consumption, continues at a rate that can be called the resting or maintenance rate.[1] The regular relationship between resting metabolic rate, or rate of

1 The commonly used term "basal" metabolic rate is not well chosen, for it incorrectly implies that there is a certain level below which the metabolic rate will not fall. This is patently wrong; for example, the metabolic rate of humans during sleep falls below what is considered the basal metabolic rate. When it comes to cold-blooded animals, the metabolic rate varies with temperature and so many other external factors that use of the term "basal" is absurd. As a result, the term "basal" is less and less used (except in studies of humans), and we shall not use it here.

Figure 6.1. Metabolic rates for mammals and birds, when plotted against body mass on logarithmic coordinates, tend to fall along a single straight line. Adapted from Benedict (1938).

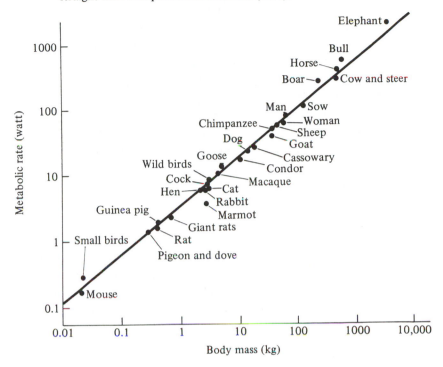

oxygen consumption, and the body size of an animal is well known to all biologists. It is striking that when the metabolic rates of birds and mammals are plotted against body mass on logarithmic coordinates, the points fall along a straight line (Figure 6.1). This line has become known as the "mouse-to-elephant curve." It has been widely discussed, and there has been much speculation about the reasons for the obvious regularity.

Most of this chapter will be devoted to further exploration of the relationship between metabolic rate and body size in various groups of animals. In later chapters we shall be concerned with the additional demands on the energy supply system caused by activity and locomotion.

Metabolic rates of mammals

Physiologists have given more attention to mammals than to any other animal group, and a tremendous amount of information is available on their metabolic rates. Such information has been collected for more than a century, and for about half of that time considerable efforts were made to fit the observations to a preconceived model that can be simply described as the "surface rule." More about that later.

We shall instead begin with the year 1932, when the distinguished physiologist Max Kleiber wrote a now famous article: "Body Size and Metabolism" (Kleiber, 1932). It appeared in a little-known journal of agricultural science, the *Hilgardia*, published by the California Agricultural Experiment Station at Davis, the institution that later became the University of California at Davis. Kleiber surveyed metabolic rates for animals ranging in size from rats to steers, a 4000-fold difference in size, from 0.15 to 679 kg. By expressing metabolic rate (P_{met}, in kcal/day)[2] as a function of body mass (M_b, in kg) in an allometric equation, Kleiber found that the best fit for his data was:

$$P_{met} = 73.3 \, M_b^{0.74} \tag{1}$$

This corresponds to a plot of the data on logarithmic coordinates that will give a straight regression line with a slope of 0.74. If the metabolic rate were a direct function of body surface area, the slope and the expected exponent in the equation would be 0.67. Kleiber's contribution has had profound influence by setting the stage for common use of allometric equations to express empirical metabolic data.

As additional observations have extended the size range and number of animals, Kleiber's estimate of the regression line that best fits the observations has essentially been confirmed. Two years after Kleiber's publication, Brody and associates (1934) included many additional species, ranging from mice to elephants, and published their well-known mouse-to-elephant curve. The slope of their regression line was 0.734, nearly identical with that of Kleiber; thus, they confirmed that the metabolic rates of mammals are not proportional to their body surface areas. Four years later, Benedict (1938) published a similar graph that again showed that in a logarithmic plot, the metabolic rates for birds and mammals fall amazingly close to a straight regression line.

---

2 The symbol P is used for power. Power (dimensions $M L^2 T^{-3}$) is defined as work or energy (dimensions $M L^2 T^{-2}$) per unit time. The rate of oxygen consumption, if considered equivalent to the energy metabolism per unit time, is therefore the metabolic power (P_{met}).

Earlier investigators often converted rates of oxygen consumption to rates of heat production by equating 1 liter of oxygen with 4.8 kcal. It has recently become increasingly common to express metabolic rates in watts (W, or J/sec), recalculated from oxygen consumption by equating 1 liter of oxygen with 20.1 kJ. The next several pages will, for historical reasons, maintain the traditional units of calories per unit time.

Kleiber has repeatedly reviewed this subject, and in his book *The Fire of Life* (1961) he discussed the meaning (or lack of it) that can be ascribed to the small numerical differences in the exponents reported by various investigators. First of all, Kleiber suggested that it would be easier to use an equation with the same units as in equation (1), with the exponent rounded off to 0.75, as follows:

$$P_{met} = 70 M_b^{0.75} \qquad (2)$$

Kleiber chose the exponent 0.75 because the numerical difference from 0.73 is insignificant, and, he said, calculations are facilitated if the rounded-off exponent is used.[3] This is quite acceptable, for the second decimal of the exponent is statistically uncertain. We should also remember that the data used to calculate the equation were collected by different investigators, by a variety of techniques, and often under different circumstances. Thus, it is usually assumed that a "true" resting metabolic rate is obtained on fasting animals; for rats this may be 12 hr after the last feeding, but for a ruminant such as the cow, probably 4 or 5 days. Because starvation causes a decrease in metabolic rate, this would require a finer distinction between "fasting" and "starving" than we can establish. It was for this reason that Brody, instead of fasting the elephants he studied, "adjusted" the values for their metabolic rates before calculating the regression equation (see Chapter 3). Because the elephant is at the extreme of the size range, that adjustment had a disproportionate effect on the regression line. Without the 30% adjustment used by Brody, his calculated slope (0.734) would have been closer to that suggested by Kleiber (0.75).

There is no reason to review in detail the tremendous amount of information that has been published on metabolic rates of mammals, for we shall not be able to establish a more "significant" slope. The material has been compiled and reviewed by many distinguished investigators, with two of the most comprehensive reviews published by Hemmingsen (1950, 1960).

Can we hope to find a significant difference between the exponents 0.75 and 0.734 as more data are accumulated in the future? Kleiber has calculated that, taking into account the coefficient of variation in the data used to calculate the regression line, no significant difference between

3 Kleiber's reason was that calculations are simpler if the logarithm of the body mass, M_b, is multiplied by 0.75. With a slide rule or a log table, this is most easily done by finding the logarithm of M_b and subtracting from it exactly one-quarter, an operation that can be done in one's head, whereas multiplication by 0.74 or 0.73 is more laborious. With the advent of inexpensive pocket calculators, this convenience has become irrelevant.

Figure 6.2. It has been suggested by Heusner (1982) that an overall regression line with a slope of 0.75 (dashed line) can be obtained as a "statistical artifact" from data that for each species fall on regression lines with slopes of 0.67 (solid lines).

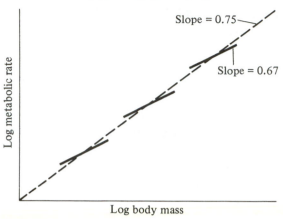

the two exponents can be established within a group of animals ranging from a minute 10-g mouse to a 16-ton super-sized elephant. In other words, the available information will not allow us to distinguish a numerical difference between slopes of 0.73 and 0.75, however much we may wish to try. On the other hand, it is overwhelmingly certain that the exponent differs from 0.67 (which would be the exponent if metabolic rate were proportional to body surface area). To establish the difference between 0.75 and 0.67 with statistical certainty requires only a 10-fold difference in body size (Kleiber, 1961, p. 212).

Is the "true" slope really 0.75?

It has been widely accepted that the slope of the metabolic regression line for mammals is 0.75 or very close to it, and most definitely not 0.67 (as the "surface rule" would suggest). How true is this? Two recent publications must be discussed in this context.

It was suggested by Heusner (1982) that the 0.75 exponent in Kleiber's equation is a statistical artifact. The reasoning is straightforward. Figure 6.2 shows the principle involved. Heusner suggested that the mass exponent for any one species is 0.67 and that the proportionality coefficient, *a*, changes as one moves from small to large animals. He examined data from seven species of mammals, ranging from 16-g mice to 922-kg cattle. A regression equation with a mass exponent of 0.67 was calculated for each species, with the result that the proportionality changed by threefold from the smallest to the largest species, from 1.91 in the mouse *Peromyscus* to 6.06 in cattle (units in watts and kilograms).

This statistical treatment gave a valid exponent of 0.67 at the level of any one species. However, comparing different species over the full size range, the result was that the proportionality coefficient had become a variable. It is simply a question of removing part of the 0.75 exponent and inserting it into *a*, or another way to account for the fact that mammals in general are not geometrically similar and are not adhering to the surface rule.

The statistical problems raised by Heusner were discussed in considerable detail by Feldman and McMahon (1983) in an article entitled "The 3/4 Mass Exponent for Energy Metabolism Is Not a Statistical Artifact." They concluded that Heusner had demonstrated that the exponent $\frac{2}{3}$ provides a better description of intraspecific variation in a set of animal data from the literature and therefore had dismissed the value of $\frac{3}{4}$ as a "statistical artifact." Using the same data, Feldman and McMahon recast the statistical analyses in mathematical equivalent form and showed that $\frac{3}{4}$ nonetheless emerges as a precise numerical estimate of the mass exponent for interspecific variation. It remains, then, that an equation with a constant proportionality coefficient, *a*, and an exponent of 0.75 is a valid statistical description of the data accumulated on the metabolic rates of mammals.

Another discussion of the validity of the body mass exponent has come from Bartels (1982). Bartels noted that the smallest mammals, especially shrews, have metabolic rates well above the common regression line for mammals. He suggested that the metabolic rates, as well as associated functions of the gas transport system in small mammals, are higher than indicated by the commonly accepted regression line. By separating out mammals between 2.4 and 100 g, he found within this group a slope of 0.23. If he extended the range from 2.4 to 260 g, the slope was 0.42, and if only animals between 260 g and 3800 kg were included, the slope was 0.76.

This treatment raises problems that are difficult to resolve at the present time. There is no doubt that shrews, as a general rule, have very high metabolic rates; they are nervous and restless and may not fit the general pattern for mammals. If a sufficient number of points for shrews are sequestered out, the inevitable result is that this group must deviate from the overall line for mammals. Second, if a regression is calculated from all available data, a sufficient number of such deviants at the extreme of the size range will have an undue effect on the overall slope. Nevertheless, it remains that shrews do deviate from the general mammalian pattern.

Unfortunately, we do not yet have sufficient understanding of the principles that underlie the slope we obtain for mammals over the full size range known to us. We can sequester some animals (in this case the shrews) that seem to deviate from the general mammalian pattern and examine what in this case appears as a secondary signal. It is interesting to consider the associated adjustments that must be required for blood, circulation, heart, and other parameters of the oxygen supply needed to sustain these high metabolic rates, but this gives little insight into the general mammalian pattern, which, based on present information, seems to fit a slope of 0.75.

Specific metabolic rate

Until now we have considered the metabolic rate of an entire animal. However, for many purposes it is convenient to compare animals of different sizes with their metabolic rates expressed per unit mass.

A great deal of confusion can arise if the same term, metabolic rate, or rate of oxygen consumption, is used to refer to the entire animal and to the unit mass. If the units are carefully stated, the problem can always be solved, but this is not always the case. To avoid confusion, the word "specific" is used to designate "per unit mass," thus referring to the metabolic rate per unit mass as "specific metabolic rate." This is in accord with the definition of the word "specific" before the name of a physical quantity, which means "divided by mass" (Royal Society, Symbols Committee, 1975, p. 10). I shall use the word "specific" in this sense, and it will be symbolized by an asterisk, as used by Weis–Fogh (1977). Thus, specific metabolic power will be written as P_{met}^*.

The specific metabolic rate (kcal day^{-1} kg^{-1}) can be obtained from equation (2) by dividing by body mass, M_b (kg), as follows:

$$P_{met}^* = \frac{P_{met}}{M_b} = \frac{70 M_b^{0.75}}{M_b} = 70 M_b^{-0.25} \tag{3}$$

This tells us that the specific metabolic rate, the metabolic rate per kilogram body mass, decreases with increasing body mass (negative exponent), and if the specific metabolic rate is plotted against body mass on logarithmic coordinates, the regression line will have the negative slope −0.25 (Figure 6.3).

Figure 6.3. The specific metabolic rate (P^*_{met}, the metabolic rate per unit body mass, M_b) decreases with increasing body size, the regression line having a slope of -0.25.

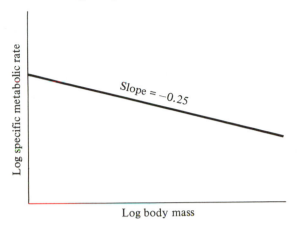

What this equation says is simply that, per unit body mass, energy metabolism is higher in a small than in a large mammal. For example, the expected specific metabolic rate of a 30-g mouse calculated from the equation will be 168.2 kcal kg^{-1} (24 hr)$^{-1}$. For a 300-kg cow it will be 16.82 kcal kg^{-1}(24 hr)$^{-1}$. In other words, for a 10 000-fold increase in body size, the specific metabolic rate will decrease to one-tenth (10 000$^{-0.25}$ = 0.1).

Of course, it is unrealistic to expect that a regression line (or an equation) will tell precisely what the specific metabolic rate will be for an animal of a given body mass. What the equation tells is the expected mean value for a "typical" mammal of the given size, and real animals will always deviate more or less from this idealized norm. If the deviations are particularly great, they probably merit further attention. If technical errors can be excluded and the observations are genuine, they may reveal interesting information. For example, it appears that seals and whales have metabolic rates about twice as high as expected for their body sizes; this deviation from the "normal" metabolic level of mammals may be related to the special problems of heat regulation and maintaining a high body temperature in cold water (Andersen, 1969). Another example: It has been found that many desert mammals, ranging from camels to small rodents, have metabolic rates substantially lower than expected from their body size. In this case the interpretation is not as straightforward; it may be related to the fact that food often is less readily available in the desert than in other environments (Yousef and Johnson, 1975).

These examples show one way in which a general equation for the metabolic rates of mammals can be useful: It gives us a standard against which to compare any given observation, be it for a seal, a camel, or a desert rat. The deviations seem to be more than random variations or noise; they appear to be secondary signals that convey a specific message.

Marsupial mammals

It is often stated that marsupials represent a more "primitive" evolutionary level than the eutherian placental mammals. However, it is a mistake to assume that functional characteristics of marsupials are primitive in the sense of being simple or inferior. The fact is that marsupials, after a separate evolutionary history of some 100 million years, represent a functional level similar to that of eutherian placental mammals, and with many striking similarities.

Support for the purported "primitive" level of marsupials has come from the observation that the usual body temperature in this group is somewhat lower than that in the "higher" eutherian mammals. A much-quoted study by C. J. Martin (1903) interpreted the lower body temperatures of marsupials to mean that these are more "cold-blooded" than eutherian mammals. I have previously emphasized that marsupials by no means are "primitive" in the sense of having inadequate temperature regulation (Schmidt–Nielsen, 1964). We shall later return to this; for the moment we shall focus only on energy metabolism of Australian marsupials.

Two recent studies have surveyed energy metabolism of a wide range of Australian marsupials (MacMillen and Nelson, 1969; Dawson and Hulbert, 1970). The former study surveyed 12 species of the family Dasyuridae, ranging in size from a 7.2-g marsupial mouse to the 5-kg Tasmanian Devil. It was found that their oxygen consumption ($\dot{V}_{O_2}$) could be described by the logarithmic equation $\dot{V}_{O_2} = 2.45\,M_b^{0.739}$ ($\dot{V}_{O_2}$ in ml O_2 hr^{-1}, and body mass, M_b, in g). Recalculated to the units we used before, heat production in kilocalories per 24 hr and body mass in kilograms, the equation will be $P_{met} = 46.52\,M_b^{0.739}$.

The other study (Dawson and Hulbert, 1970), which included the large kangaroos, had animals ranging in size from 9 g to 54 kg, and although the number of species was only eight, the size range was 6000-fold. The regression equation for metabolic rate (P_{met}, in kcal day^{-1}) versus body

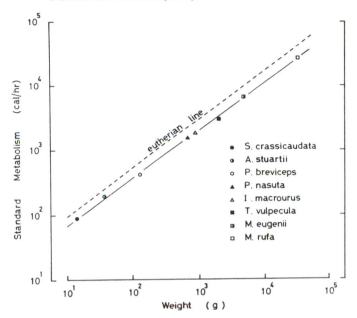

Figure 6.4. Standard metabolism–body-size relationship in marsupials. The eutherian line is drawn from the data of Kleiber (1961). From Dawson and Hulbert (1970).

mass (M_b, in kg) was similar to the preceding: $P_{met} = 48.6\,M_b^{0.737}$ (Figure 6.4).

The slopes as well as the proportionality coefficients are virtually identical in the two equations. The two studies thus agree that the metabolic rates of marsupials vary with body size in the same way as for placental mammals (the slopes are virtually identical); however, the level of the marsupial metabolic rates (as expressed by the proportionality coefficients) is 30% lower than that for eutherian placental mammals. We shall return to this difference later in our discussion of mammalian body temperatures.

Birds

A tremendous amount of published data on energy metabolism in birds was compiled and reviewed by Lasiewski and Dawson (1967). For reasons that could be disputed, but with which I happen to agree, they separated the birds into two groups: one containing only passerines (sparrows, finches, starlings, crows, etc.) and the other containing all other birds. They then calculated metabolic regression lines for each of the the two groups separately. Their equations (P_{met}, in kcal day^{-1}, and M_b, in kg) were, for nonpasserine birds, $P_{met} = 78.3 M_b^{0.723}$, and for passerine birds, $P_{met} = 129 M_b^{0.724}$.

Let us first discuss the reasons for separating passerine and non-passerine birds. The former range in size from a 6-g finch to nearly 1 kg for the raven; the latter range from a 3-g hummingbird to the 100-kg ostrich. The total size range for the nonpasserine birds is therefore more than 100 times greater than that for the passerines. For the moment, let us accept that a separation is valid and that the passerines indeed have a higher metabolic level than the nonpasserines. The passerines are located only at the lower end of the graph in Figure 6.5. If we include these with all the other birds, the large number of high values for passerines in the small-body size range will tend to pull the left end of the regression line (the dashed line) upward. If this is done, the equation for all birds, passerines and nonpasserines combined, is $P_{met} = 86.4 M_b^{0.668}$.

In this case we cannot cogently argue that one procedure is right and the other wrong. We are merely discussing a regression line that describes empirical data, and in either case the equation is a valid description of the data we include, calculated as the least-squares linear-regression line in a logarithmic plot. It is interesting to note, however, that if all birds are included, the regression line has a slope that describes a perfect proportionality to body surface (provided that the surface area for birds is indeed proportional to the two-third power of the body mass). On the other hand, if passerines are treated as a separate group, one can say that all birds have metabolic regression lines virtually identical in slope with those of mammals, that nonpasserine birds are numerically at the same level as placental mammals, and that passerine birds, on the whole, are at a level nearly twice as high as mammals and other birds. I believe that these generalizations are now commonly accepted.

The preceding equations refer to fasting birds at rest, but one variable was not considered: the time of day. However, birds have a profound daily rhythm, and if the data are reexamined in this light, an interesting fact is revealed (Aschoff and Pohl, 1970a, 1970b). Let the normal activity period for a bird be designated as the alpha (α) period, and the normal resting period as the rho (ρ) period. For a diurnal bird, the α period occurs during the daylight hours, and the ρ period at night; for a nocturnal bird the situation is the reverse.

Aschoff and Pohl examined available metabolic information on birds and separated observations made during the α and ρ periods. Determinations during the α period were made on birds at rest and in the dark, but

Figure 6.5. Comparison of regression lines for passerine birds, nonpasserine birds, and all birds. From Lasiewski and Dawson (1967).

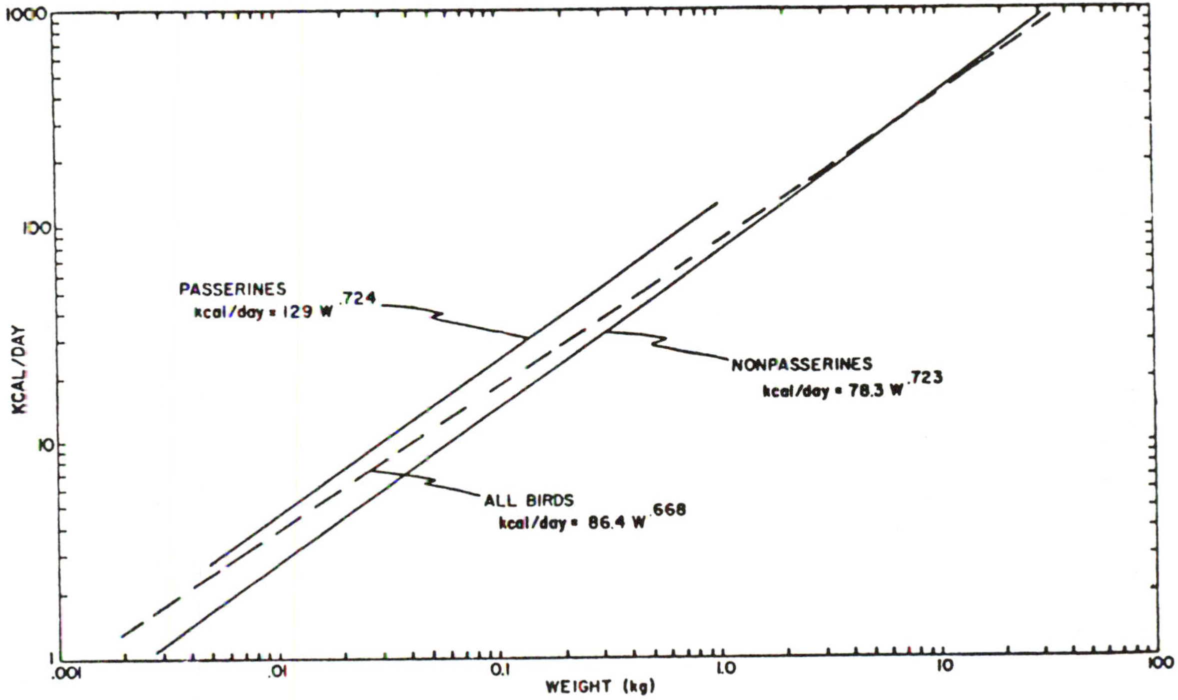

Figure 6.6. Resting metabolic rate as a function of body size in passerine and nonpasserine birds. The solid lines are regression lines for values obtained during normal activity (α) and rest (ρ), as marked on the graph. The dashed lines represent regression equations from Lasiewski and Dawson (1967). From Aschoff and Pohl (1970*b*).

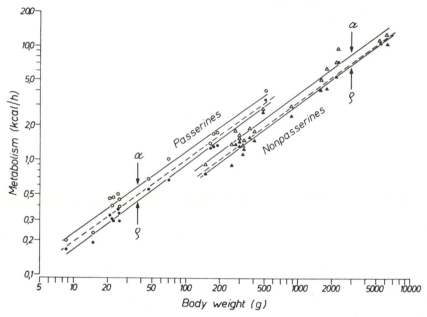

they were nevertheless about 25% higher than the ρ values. This is expressed in Figure 6.6, which treats passerine and nonpasserine birds separately. We should note that α and ρ values were obtained under similar conditions on birds resting and in darkness, the only difference being whether they were obtained during the normal activity period or rest period for the birds. The corresponding equations for these data, given in the units used before (kcal day^{-1}, and body mass in kg), are shown in Table 6.1.

The inevitable question arises: What is the true resting metabolic rate for birds? The question is unanswerable; we can only say that we must carefully define the conditions for the data under discussion and consider some variables that at times may be overlooked, such as, in this case, diurnal rhythms.

Reptiles

The metabolic level for cold-blooded animals is poorly defined. It changes drastically with changing temperature and with numerous

Table 6.1. Regression equations for resting metabolic rates of birds (P_{met}, kcal day^{-1}) versus body mass (M_b, kg) obtained during the normal activity (α) period and rest (ρ) period (Aschoff and Pohl, 1970*b*).

	Metabolic rates
Nonpasserine birds:	
α period	$P_{met} = 91.0\,M_b^{0.729}$
ρ period	$P_{met} = 73.5\,M_b^{0.734}$
Passerine birds:	
α period	$P_{met} = 140.9\,M_b^{0.704}$
ρ period	$P_{met} = 114.8\,M_b^{0.726}$

other factors such as feeding or merely the presence of food, light, season, and, in particular, the previous thermal history of the animal. It is therefore impossible to designate one particular level as being the normal resting metabolic level for a cold-blooded animal. The best we can do is to specify the conditions as precisely as possible, with particular attention given to temperature.

This is not the place for an extensive review of the metabolic rates of reptiles or other cold-blooded animals; a few examples will suffice to show the general trends.

A compilation by Bartholomew and Tucker (1964) reviewed their own results as well as other published results for lizards ranging in size from 2 g to 4.4 kg. The data (obtained at 30°C) yielded the equation $P_{met} = 6.84\,M_b^{0.62}$, with 95% confidence limits on the exponent of ±0.08 [recalculated to kcal (24 hr)$^{-1}$, body mass in kg].

A more recent summary (Bennett and Dawson, 1976) of data for 24 species of lizards, ranging from 0.001 to 7 kg in size, gave the equation (units as before)

$$P_{met} = 7.81\,M_b^{0.83 \pm 0.01}$$

The exponent is significantly different, which merely shows that the numerical value of a calculated exponent reflects the material that happens to be available, rather than the biological meaning of a number with a given statistical significance.

This becomes clear from a study of metabolic rates of tropical snakes (Galvão et al., 1965), which raises problems we cannot easily resolve. The study involved 50 adult snakes of 18 different species, ranging in

Table 6.2. Metabolic rates [P_{met}, in kcal $(24\,hr)^{-1}$] of tropical snakes over a 2000-fold range in body mass (M_b, kg), from 0.011 to 22 kg (Galvão et al., 1965).

Snakes	Metabolic rates
Colubridae	$P_{met} = 4.390\,M_b^{0.98 \pm 0.04}$
Boidae	$P_{met} = 1.788\,M_b^{1.09 \pm 0.09}$
All snakes	$P_{met} = 3.102\,M_b^{0.86 \pm 0.03}$

size from 11 g to 22 kg. The snakes belonged to two different families, the Colubridae and the Boidae. The calculated regression lines are shown in Table 6.2.

First of all, we notice that the metabolic rate for a colubrid snake of unit mass (1 kg) is somewhat less than that for a lizard of unit mass (as discussed earlier) and that a boid snake of unit mass has a metabolic rate less than one-half of that for a colubrid snake. The exponents (i.e., the slopes of the regression lines) are close to unity for both groups. This means that among these snakes, metabolic rate is proportional to body mass. This observation differs considerably from what we have seen for other vertebrate groups.

A curious circumstance is revealed when all the snakes from the full size range of the two groups are pooled. In that event, the regression line (Figure 6.7) has a slope of 0.86, that is, lower than in either of the two groups separately. The reason is evident from the graph: Each of the two groups is represented by a relatively steep slope, but because the boid snakes are concentrated on the right side of the graph and are at a lower level than the colubrid snakes, the overall regression line for all the snakes combined has a lower slope than that for either of the two component groups.

This result of pooling a large number of data is the same as we observed for the birds, only in this case the effect is more pronounced. We are left with a dilemma: the question whether it is correct to separate the groups or to treat them together as "all snakes." There is no ready answer. In either case we have specified the material used for the calculation, and either set of regression lines is arithmetically correct. But when it comes to interpretation of their biological meanings, the dilemma is not so easily resolved.

What slope one finds for a metabolic regression line for snakes depends on how one views the information at hand, and this amplifies

Figure 6.7. Metabolic rates for 50 specimens of tropical snakes, measured at 20°C. From Galvão et al. (1965).

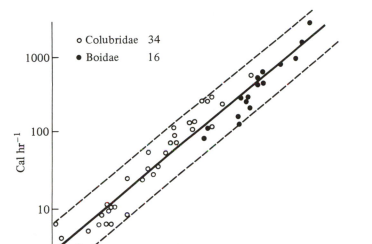

the difficulties in finding a definite answer to what the slope is. The data that Galvão used to show that the slopes for colubrid snakes as well as boid snakes were close to 1.0 were included in an analysis by Dmi'el (1972), who arrived at a very different slope: 0.6. This example is instructive: It reveals a slope of about 1.0 when a small homogeneous group of relatively narrow size range is examined; if the material is included in a broader grouping over a wider size range, a different slope appears to apply. There are many similar examples, especially among groups of invertebrates; the meaning one may wish to attach to them depends on one's viewpoint in choosing the data base.

Amphibians and fish

It was difficult to establish a reasonably reliable slope for the metabolic relationship to body size for reptiles, and we meet with similar difficulties for the more aquatic vertebrates: amphibians and fish.

The metabolic rate of cold-blooded aquatic animals is not only sensitive to temperature and other variables that we know about but also shows a strong dependence on the oxygen supply. This additional variable makes it even more difficult to establish a "standard" resting metabolic rate than it is for terrestrial vertebrates that breathe air with a constant oxygen content. Therefore, determination of an exact slope for the regression lines is even more uncertain for aquatic animals than it is for the higher vertebrates.

Anuran amphibians (frogs and toads) were studied by Hutchison and associates (1968), who examined pulmonary as well as cutaneous gas exchange in 20 species. They included temperature as a variable and obtained information from 5°C to 25°C. The slopes of the regression lines between metabolic rate and body mass ranged from 0.59 to 0.94, with an arithmetic mean for the observed exponents of 0.71.

Urodele amphibians (salamanders) were studied by Ultsch (1974). In this case, only three species were examined, but one of them, *Siren lacertina,* covered a tremendous size range, from 0.36 g to 1310 g. This range, more than 3000-fold, permitted the calculation of a reasonably precise exponent, which was 0.66. This is virtually a complete proportionality to body surface area, and in fact the observed regression line for surface area to body mass in these animals had a slope of 0.65 (with a *k* value of 10.5). The metabolic regression equation at 25°C was $P_{met} = 0.79\,M_b^{0.66}$ (kcal day^{-1} and kg).

What we can conclude from this information is perhaps the following: Amphibians have metabolic-rate–body-size relationships that in the usual logarithmic plot have slopes not too different from those for the higher vertebrates. Whether the slopes are closer to 0.67 (a surface relationship) or to 0.75 (as for mammals and birds) is difficult to say. What can be stated with fair certainty, however, is that the relationship consistently differs from proportionality to body size (slope = 1.0).

When it comes to fish, it is even more difficult to determine a general slope for the metabolic-rate–body-size regression lines. There are, of course, the same difficulties as with other cold-blooded animals: temperature, thermal history of the animal, oxygen content of the water, feeding and nutritional state, light or dark, and so on. To the light parameter we must also add the light–dark cycle (day length), which profoundly influences many physiological variables. No wonder that different authors obtain different and at times conflicting results. We should also remember that in addition to many other possibilities for discrepancies, there may be true differences between species, or even between subspecies and local races of the same species.

The upshot of the whole thing is uncertainty, the conclusion being that, at the present time, it is difficult to make a definitive analysis of the scaling of metabolic rate for fish relative to body size.

In addition to many other variables, the activity of the animal enters into the picture. Thus, speckled trout (*Salvelinus fontinalis*) studied over a temperature range from 5 to 20°C gave slopes from 0.80 to 0.86 for "standard" metabolism, whereas in active fish the slopes varied from 0.75 to 0.94, the steepest slope applying to the lowest temperature (Job, 1955). Much of the rather extensive literature was reviewed by Fry (1957), and some more recent contributions were listed by Ultsch (1973). There is no reason to add further listings here, but we shall return to the subject in a later chapter, when we discuss the scaling of respiratory organs, because a considerable amount of information is available for the scaling of the gills of fish in relation to their need for oxygen.

Invertebrates

When it comes to invertebrate animals, it is even more difficult to arrive at a coherent picture. To begin with, it seems unreasonable to expect that animals as different as sessile coelenterates (e.g., sea anemones) and highly active crustaceans (e.g., swimming crabs) should have similar metabolic rates. Within groups of similar animals, especially if the range is reduced to a single genus or species, we can obtain more meaningful information about the size dependence of the metabolic rate.

This is not the place to compile a list from the immense amount of scattered information available in the literature. One such listing, giving metabolic rate and the slope of the body size regression line for some 200 invertebrates (Altman and Dittmer, 1968), shows quite clearly how heterogeneous the situation is: The listed slopes vary, without any discernible regularity, from less than 0.67 to over 1.0.

Aside from compiled lists, the subject has been covered by several broad and competent reviews, some notable examples being those by Hemmingsen (1950, 1960) and Zeuthen (1947, 1953). Zeuthen made a meritorious contribution, not only by his extensive review but also by a high-quality experimental study that included observations on a broad range of very small organisms, down to microscopic size. Hemmingsen's two monographs are perhaps the most monumental reviews of the entire subject of metabolic rate and body size; they of course include careful discussions of Zeuthen's contributions.

Hemmingsen pointed out that although few of the individually determined regression lines had slopes of 0.75, most of the deviating short-range lines were located close to a "standard line" with a slope of 0.75 that stretched over a much wider range (Figure 6.8). According to Hemmingsen (1960, p. 33), this fact confers upon the standard line the

Figure 6.8. Relationship of energy metabolism to body mass as found by various authors within shorter poikilotherm ranges as compared with the standard relationship for the entire poikilotherm range (dotted line). 20°C. From Hemmingsen (1960).

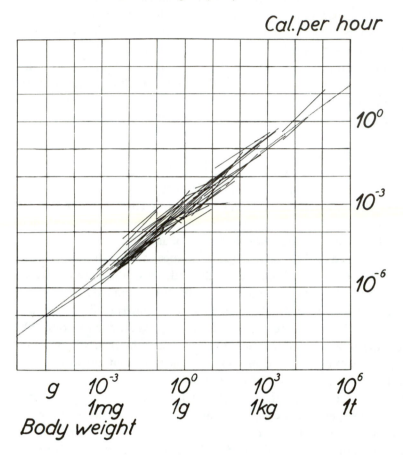

distinction of being representative of the whole range of poikilotherm or cold-blooded animals. It may be justified to accept the overall slope of 0.75, as proposed by Hemmingsen, if only we remember that there is room for much variation. Within narrow ranges of similar animals, for single genera or species, the slopes may give an entirely different picture.

At the present time, the information on invertebrate metabolic rates is not easily analyzed in the light of scaling principles. We shall therefore return to a more detailed discussion of the warm-blooded vertebrates, birds and mammals, that have more stable metabolic rates and for which a resting maintenance rate can be established with reasonable certainty.

7

Warm-blooded vertebrates: What do metabolic regression equations mean?

Let us return to the warm-blooded vertebrates, birds and mammals. We have seen that for these the empirical relationship between metabolic rate and body size is known with much greater certainty than for cold-blooded vertebrates and invertebrates, and the reason is undoubtedly related to their relatively constant body temperature. To understand the regular relationship to body size, we should examine those groups that are best known, also in regard to related physiological information.

Body temperature

First of all, the regularity of metabolic rates in warm-blooded vertebrates is undoubtedly related to their relatively constant body temperatures. This eliminates temperature as a variable here, whereas in cold-blooded animals it causes innumerable difficulties in determining a definite metabolic rate, both for the short term and for the long term (acute effects of a temperature change as well as long-term acclimatization).

How constant is the body temperature of birds and mammals? For the moment, let us disregard the small number of mammals and even smaller number of birds that in connection with periods of torpor or hibernation can undergo profound decreases in body temperature. Here we are concerned with the normal or usual temperature of active animals.

Do small and large mammals have similar temperatures, or is the body temperature of mammals related to body size? This question was discussed by Morrison and Ryser (1952), who examined published material supplemented with a considerable number of observations of their own, especially on smaller animals. They divided the animals into size groups ranging from 1 to 10 g, 10 to 100 g, and so on. The temperature ranges

Table 7.1. Body temperatures of eutherian mammals arranged according to body size (Morrison and Ryser, 1952).

Body mass (kg)	Number of species	Body temperature (°C)	
		Range	Mean
0.001–0.01	2	37.8–38.0	37.9
0.01–0.1	11	35.8–40.4	37.8
0.1–1.0	12	35.8–39.5	37.8
1.0–10	17	36.4–39.5	38.0
10–100	8	36.0–39.5	37.9
100–1000	6	36.4–39.5	37.8
1000–10 000	2	35.9–36.1	36.0
10 000–100 000	4	36.5–37.5	37.1

and mean values listed in Table 7.1 show no clear correlation or obvious trend in relation to size. The range of temperatures for all groups is roughly between 36 and 40°C, with an overall mean close to 38°C.

Could climatic region be of any importance, so that arctic and tropical mammals have different temperatures? Material compiled by Irving and Krog (1954) did not reveal any significant differences; the mean body temperature for 19 arctic and subarctic mammals in the resting condition was 38.3°C, slightly but not significantly higher than the mean for the data presented by Morrison.

Body temperatures of birds have been measured by many investigators, and extensive tabulations have been published. It has been claimed that small birds tend to have higher body temperatures than larger birds, but this claim cannot be accepted without some reservations. Many of the published temperatures for small birds were measured with mercury thermometers, but the metabolic rate of small birds is so high that body temperature can increase by one or two degrees in one minute. Even when a resting bird is rapidly removed from its cage and a thermometer or a probe is inserted, the temperature of the hand-held specimen may easily be misleadingly high before a reading is obtained. Such measurements are, of course, highly suspect.

It is also necessary to consider that birds have a normal diurnal temperature variation of a couple of degrees. This led Calder and King (1974) to review earlier compilations. Their conclusion was that body temperatures of birds, if measured with minimal disturbance and during rest, appear to be independent of body size. They suggested an average of about 40°C ± 1.5°C, that is, about 2°C higher than the typical body temperature for mammals.

I am not aware of any extensive compilations of body temperatures of the more "primitive" mammals. It appears that marsupials have body temperatures characteristically about 2°C below those for eutherian placental mammals, about 36°C. Marsupial temperatures have been recorded by Dawson and Hulbert (1970) and by MacMillen and Nelson (1969). The former reported body temperatures of about 36°C or some-what lower; the latter (for dasyurid marsupials only) reported tempera-tures closer to 37°C. There was no clear trend with respect to body size among the marsupials; some of the smallest had lower temperatures than the mean, but other small species had temperatures somewhat higher than the mean.

The most primitive living mammals are the egg-laying monotremes, represented by the spiny anteater or echidna (*Tachyglossus*) and the duck-billed platypus (*Ornithorhynchus*). Both these animals maintain a normal body temperature of about 30 to 31°C. This does not imply that their temperature regulation somehow is "primitive" or that they are intermediate between warm-blooded higher mammals and cold-blooded lower vertebrates. On the contrary, both echidna and platypus are com-petent temperature regulators that can maintain their body temperature at a nearly constant level, even at ambient temperatures down to freezing (Schmidt–Nielsen et al., 1966). In this regard they are by no means "primitive," although their tolerance to high temperatures is limited.

In summary, each major group of the higher vertebrates appears to maintain, within ±1°C or so, its own characteristic resting body tem-perature, which for birds is 40°C, for eutherian mammals 38°C, for mar-supials 36°C, and for monotremes 30°C. Although the major groups differ from each other, within each group the resting body temperature is maintained rather constant, and without any obvious relation to body size. Small and large mammals, on the whole, have body temperatures at the same level.

The surface law

When small and large mammals maintain the same body tem-perature, why is the relative rate of heat production higher in the small animal? The answer is well known to all biologists: The small animal has, relative to its mass, a larger body surface. Heat loss takes place from the surface, and in order to keep warm, an animal must produce heat at a rate equal to the loss.

Table 7.2. The earliest study that related metabolic rate to body surface area rather than to body mass was a study of dogs. Although there was better correlation with surface area than with body mass, it is now clear that body surface area per se does not determine metabolic rate. Nevertheless, body surface area constitutes a constraint on what is possible in the design of a well-functioning animal (data from Rubner, 1883).

Body mass (kg)	Body surface (cm^2)	Surface: mass ratio (cm^2/kg)	Metabolic rate	
			Per mass [kcal (kg day)$^{-1}$]	Per surface area [kcal (m^2 day)$^{-1}$]
31.20	10 750	344	35.68	1036
24.00	8805	366	40.91	1112
19.80	7500	379	45.87	1207
18.20	7662	421	46.20	1097
9.61	5286	550	65.16	1183
6.50	3724	573	66.07	1153
3.19	2423	726	88.07	1212

This simple relationship has been well understood since the early part of the last century. In 1839, Sarrus, a professor of mathematics in Strasbourg, and Rameaux, a doctor of medicine, wrote a thesis on this subject that was read to the Royal Academy of Medicine in Paris. They emphasized that the heat loss from a warm-blooded animal must be roughly proportional to its free surface, and because a small animal has a larger relative surface, it must also have a higher relative rate of heat production to keep up with the heat loss.

This reasoning is fundamentally sound, and it was adopted by Bergmann (1847), who formulated the well-known Bergmann's rule. This rule states that warm-blooded animals in colder climates are of larger body size (i.e., have smaller relative external surface areas) than their relatives in warmer climates. The validity of this rule has been questioned and has caused much controversy.

The first experimental examination of the relationship between heat production and surface area was made by Rubner a century ago (1883). He studied heat production (measured as oxygen consumption) in dogs of various sizes and found that the smaller the dog, the higher its heat production per kilogram body mass (Table 7.2). However, if heat production was calculated per body surface area (last column), there was a nearly constant ratio between heat production and body surface of the dogs. Rubner thought that this finding confirmed that heat production is adjusted to the animal's needs to keep warm, for heat loss takes place from the surface. He also explicitly stated that heat production is due to stimulation of receptors in the skin, which in turn act on the cells of the metabolizing tissues. This view is now considered erroneous.

Rubner's findings led to general acceptance of a "surface law" and to extensive use of body surface as a base of reference for metabolic rates, notably in clinical medicine. The surface law is deceptively attractive, but it is not a law in the sense of physical laws, such as the law of gravitation, for example, and it could better be described as a "surface rule."

Could it be that warm-blooded animals obey a surface rule (although we thus far have assumed that the metabolic rate in that event should be related to body size with the exponent 0.67, rather than 0.75 as found for mammals)? However, animals are not geometrically similar or isometric. If their relative surfaces were to vary with body mass with an exponent different from 0.67, perhaps the metabolic rate (heat production) could still be surface-related?

The surface area of a sphere of volume V is $4.836\,V^{0.67}$, and this applies to a sphere of any size, in any consistent system of units (cm^2 and cm^3, or square feet and cubic feet, etc.). Similarly, the surface area of any cube is related to its volume by $6\,V^{0.67}$. For any other geometrically similar or isometric bodies, the exponent will be the same, 0.67, and only the preceding coefficient will differ. The sphere has the smallest possible surface area for its volume, and any change in its shape will increase the coefficient to a value higher than that for the sphere (4.836).

Determinations of the surface areas of animals have received a great deal of attention in connection with studies of metabolic rate and heat exchange, and many methods for measuring surface areas have been devised. Direct methods of measurement are complex and cumbersome, and it would therefore be convenient if a calculation of surface area could be based on a simpler approach. Such attempts have followed the lead of Meeh, who in 1879 suggested that the surface area of mammals (S) could be expressed by the equation $S = k\,M_b^{2/3}$.

The body mass, M_b, is easily determined by weighing, and if k turns out to be a constant, we have a simple method for estimating the surface area of any mammal. Meeh made measurements of 16 humans by covering the body, section by section, and when possible in cylindrical shapes, with strips of millimeter graph paper. From these measurements he suggested that for humans, $k = 11.2$.

There is a considerable literature concerned with empirical determination of k values for various animal species. How accurate are these determinations? Many methods have been used for measuring surface areas. An animal can be skinned and the area of the skin determined, but how does one know how much to stretch the skin? The animal can be divided into a number of cylinders and cones, and the area of each determined separately. The animal can be covered with paper and the area of paper afterward determined, either by planimetry or by weighing the paper. The surface area for a cow has been determined by covering the animal with an ink roller, counting the number of revolutions of the roller. This works well for a large animal, but it would be difficult for a small animal such as a rat (Brody and Elting, 1926). Claims for high accuracy are not necessarily meaningful. Voit (1930) made duplicate determinations on skins of rabbits and found differences of less than 4%; he therefore concluded that the error was less than ±2%. Anyone who has handled the loose and stretchable skin of a rabbit will know that pulling a bit more or a bit less probably makes a difference 10-fold the maximum error claimed by Voit.

It is not easy to evaluate which method is "best." What matters at the moment is that relatively small differences, say 10 or 20% in the k value, probably are meaningless. Various authors disagree, even when only one species is involved. For example, mean k values for the laboratory rat determined by various authors range from 7.15 to 11.6, with the extremes ranging from 6.6 to 13.0 (Altman and Dittmer, 1964, p. 121). This implies that one rat should have twice as large a surface area as another of the same weight!

When a standard set of tables can report a twofold difference in various k values for one single species, can smaller differences between different species be meaningful? Let us examine Table 7.3, which lists k for a wide variety of vertebrates. If we disregard values that differ by less than 20% from a value of 10.0, there are only a few animals that stand out. One is the hedgehog, with a k of 7.5. This animal, with its rounded form and short legs, is about as close to having a spherical shape as any mammal, and a low k may be realistic. On the other hand, an animal as "round" as a pig has a k of 9.0, the same as for cattle. The bat, with a k of 57.5, is far outside the range for other mammals; this is because the skin membranes of the wings are included.

Another high k, not as high as that for the bat, but still far outside the usual mammalian range, has been reported for an Australian marsupial, the sugar glider (*Petaurus breviceps*) (Dawson and Hulbert, 1970). Other marsupials have k values similar to those for eutherian mammals, but the k for the sugar glider is 25.7. This is because of the gliding membranes,

Table 7.3. Values of the Meeh coefficient, k, in the equation $S = k M_b^{0.67}$ (S and M_b in cm^2 and g, or dm^2 and kg) (from Benedict, 1934).

Animal	k
Mouse	9.0
Rat	9.1 (9.13)[a]
Cat	10.0
Guinea pig	9.0
Rabbit	9.75 (12.88)
Dog (over 4 kg)	11.2 (11.2)
Dog (under 4 kg)	10.1
Sheep	8.4
Swine	9.0
Cattle	9.0
Horse	10.0
Human	11.0 (12.3)
Monkey	11.8
Sloth	10.4
Porcupine	10.8
Marmot	9.3
Hedgehog	7.5
Bat	57.5
Bird	10.0 (10.45)
Frog	10.6 (9.9)
Fish	10.0
Turtle	10.0
Snake	12.5

[a] Values in parentheses are those given by Rubner (1883) for six of the same animal species.

which more than double the total surface area of the animal. In birds, on the other hand, the wing surfaces are dead feathers and are not included; the skin area of the wing itself is quite small. Additional determinations on birds have given k values similar to those for mammals. Thus, there are no great differences in surface areas between birds and mammals.

It might be expected that snakes, with their long, slender bodies, should have k values substantially higher than the listed 12.5. However, snakes have no limbs, and the slender limbs of birds and mammals contribute substantially to their total surface areas.

In the context of heat balance, the desire to refine the measurement of surface area is a rather fruitless pursuit, for it is uncertain just what is meant by surface area. Does the "true" surface of an animal include the skin area between the legs that is not exposed to the outside? Does it include the ears, and if so, both sides? And so on. In view of these questions, which mean an uncertainty of perhaps 20%, little is gained by technical refinements, and it has become increasingly common to estimate the approximate surface area for mammals from Meeh's equation, using $k = 10.0$. This simplification is valid for many purposes, but in reality the exposed surface area of a living animal changes radically with its posture, whether it curls up or stretches out, how it holds its ears, and so on. Also, if we consider animals of more unusual shapes, such as bats or flying squirrels, the simplified Meeh equation obviously does not apply.

The futility of trying to determine k with the highest possible precision was fully realized by Benedict (1934). He quoted k values for the rat ranging from 7.47 to 11.6 and considered it incomprehensible that the outer shape of one species could vary that much. He correctly ascribed the differences to the measuring methods and suggested that all values should be close to 10. He considered it useless to argue whether a precise constant is more or less valid, and he suggested that a k of 10.0 ± 1.0 would serve for most purposes. This has been thoroughly discussed by Kleiber (1961, pp. 181ff), and we shall not discuss further refinements here.

Isometric or not?

The preceding discussion suggests that the body surface area of higher vertebrates is uniformly related to the 0.67 exponent of body volume. This would be true if the animals were geometrically similar, but this most certainly is not so. Does this mean that surface area varies with body volume with an exponent different from 0.67? From the preceding it seems that this is not the case, and this viewpoint was adopted in Hemmingsen's monumental monograph (1960) (Figure 7.1).

Nevertheless, the contention that body surface is scaled exactly in proportion to body volume to the power 0.67 was disputed by McMahon (1973), who stated that Hemmingsen's data would equally well permit a line with a slope 0.63. The reason for suggesting this particular exponent is based on a theoretical analysis of the scaling problems. The consequences of increased size in regard to mechanical failure are so important that they require a detailed discussion, for they seem to provide the first sound theoretical approach toward explaining the metabolic-rate–body-size relationship with the exponent 0.75. Although this exponent has

Figure 7.1. Relationship of body surface to body mass in vertebrates. The points surrounded by a circle represent beech trees. From Hemmingsen (1960).

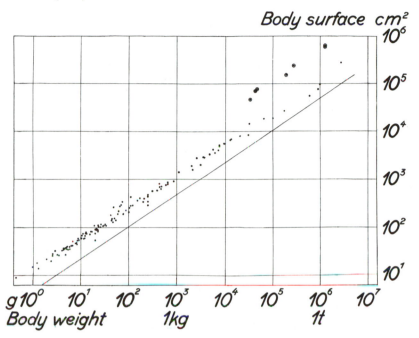

been much discussed, earlier discussions often had a certain metaphysical quality that is absent from McMahon's analysis. We shall therefore carefully discuss his considerations.

McMahon's model

Let us return to the metabolic rates for the warm-blooded vertebrates, birds and mammals. For these, a distinct resting metabolic rate can be determined with reasonable accuracy, and there is little doubt that the regression lines for their metabolic rates versus body size have slopes very close to 0.75, not to 0.67.

Can we arrive at a rational explanation for this particular slope, which definitely differs from the simple model of a surface relationship? Until recently, the attempts at explaining the deviation from a surface-related exponent have been rather unsatisfactory. It has been clear that, for a variety of reasons, the scaling of metabolic rate to body size cannot deviate too much from a surface relationship, but why should it deviate at all? Most of the explanations that have been proposed have been unsatisfactory and have had a somewhat metaphysical quality. Thus, Hemmingsen's impressive monograph (1960) ends with the statement that there are reasons to believe that the slope of 0.75 is adaptive and results "perhaps from a struggle between proportionality of metabolism to body weight and proportionality to surface functions." Seen in this light, an exponent of 0.75 should be a "compromise" between the "attempts" of the larger animal to keep its metabolism proportional to its mass and the forces that for reasons of physical necessity (e.g., heat loss) compel it to approach a surface-related level.

More recently, however, McMahon (1973) presented a model suggesting that the exponent 0.75 can be derived from theoretical considerations. As far as I know, this is the first rational attempt at referring the empirical exponent of 0.75 back to basic principles. The model deserves a careful discussion.

McMahon took as his point of departure the mechanical structure of the animal and analyzed the supporting columns (the skeleton) in the light of basic engineering principles.

Consider a tall, slender cylindrical column and the proportions necessary to keep it from failing when loaded by a force F, representing the total weight of the column acting at its center of mass. If the load exceeds the strength of the column, it will fail by rupture in compression. However, if the column is slender, it may fail in what is known as elastic buckling. If the vertical column is subjected to a small lateral displacement, the elastic forces of the bent column will return it to the vertical. However, if the weight F exceeds the elastic forces, the column will fail in elastic buckling. This type of buckling applies when the column is slender enough, and in that case the critical length (l_{cr}) for buckling is related to the diameter (d) by

$$l_{cr} = k \left(\frac{E}{\rho} \right)^{1/3} d^{2/3}$$

where E is Young's elastic modulus for the material and ρ is its density. If these are considered constant, we have

$$l_{cr}^3 \propto d^2$$

If the solid cylinder is made hollow, and if the thickness of the wall is proportional to the diameter, the same equation applies to the critical height, except for a change in the constant k. Thus, for a slender column exposed to bending forces, elastic buckling requires that the height should go as the diameter to the power $\frac{2}{3}$.

McMahon then argued that animals cannot remain geometrically similar (isometric) as their sizes increase, because in that event the cross-sectional area of their limbs would increase only as the square of the characteristic linear dimension l, although they would have to support a weight that would increase as l^3.

When a quadruped is standing at rest, its limbs are exposed primarily to buckling loads, whereas the vertebral column must withstand bending loads. When the animal runs, and the limbs provide the propulsive effort, the situation is reversed: The limbs are supporting bending loads, and the vertebral column receives an end thrust and thus a buckling load. Thus, all proportions of an animal should change with size in the same way.

Consideration of elastic criteria now leads to the following model. If w_b is the body weight, and the weight of any limb is a specified fraction of w_b, then $w_b \propto l d^2$. Elastic criteria require that $l^3 \propto d^2$; we therefore obtain $l \propto w_b^{1/4}$ and $d \propto w_b^{3/8}$.

McMahon presented empirical data to support his model of elastic similarity. One striking case concerned Brody's measurements on 3000 holstein cattle (1945). The data for chest girth G and the height at the withers H showed $G \propto w_b^{0.36}$ ($w_b^{0.37}$ predicted) and $H \propto w_b^{0.24}$ ($w_b^{0.25}$ predicted).

McMahon later presented additional anatomical material in support of his model of elastic similarity (1975b), but we shall restrict our discussion to the implications for metabolic rate.

We can now turn to the question how considerations of elastic similarity might affect metabolic rate. For this purpose, consider the power developed by the contraction of a muscle.

The work (W) performed by a contracting muscle is the product of the force of contraction and the shortening distance (Δl). We thus have

$$W \propto \sigma A \Delta l$$

where σ is the tensile stress (force per unit area) and A is the cross-sectional area of the muscle.

The power developed by the muscle, the work per unit time, will then be

$$P \propto \sigma A \frac{\Delta l}{\Delta t}$$

Evidence from muscle physiology indicates that maximum tensile stress, σ_{max}, is a scale-independent quantity that is of the order of a few kilograms per square centimeter. Also, the speed of shortening, $\Delta l/\Delta t$, appears to be a constant from species to species. This follows directly from the observation that as the total length, *l*, of homologous muscle from different animals increases, the speed of contraction, *t*, decreases in the same proportion. With σ and $\Delta l/\Delta t$ taken as constants, the power output of a particular muscle depends only on its cross-sectional area. This area (A), according to the criteria of elastic similarity, is proportional to d^2, and hence $P_{max} \propto d^2 \propto (M_b^{3/8})^2 = M_b^{0.75}$.

Let us accept that this power output applies to any particular muscle; it must therefore also apply to all metabolic variables involved in supplying the entire muscular system with energy and oxygen. This is precisely the statement we are looking for: The entire metabolic system should be scaled to body size according to $M_b^{0.75}$.

The consequence of the model based on criteria of elastic similarity thus requires that power output and related variables (respiratory exchange, blood circulation, etc.) be scaled according to body size to the power 0.75 and that physiological frequencies (such as heart rate, respiratory frequency, etc.) must scale as body size to the power -0.25. This model is in agreement with a large body of empirical material on the scaling of physiological variables that we shall discuss in the following chapters.

Gravitational effects as an explanation?

Before we turn to other animal groups, there is one suggested "explanation" for the 0.75 mass exponent that deserves discussion. This exponent could be derived from the summation of two sets of relationships, one surface-related (exponent 0.67) and the other with a higher exponent. The possibility of a summation of several component processes was in fact suggested many years ago by Kleiber.

Let us see how this could come about. In addition to the energy demands for maintenance and the functioning of all organs and tissues, land animals are under the constant effect of the earth's gravitational field. It may seem that the constant gravitational force should not affect an animal that is supported by a solid substratum, but it has been amply documented than an artificial increase in a gravitational field increases the energy metabolism (Smith, 1976, 1978). Assume that the maintenance metabolism is directly surface-related (exponent 0.67) and that the work against the gravitational field is directly mass-related (giving the exponent 1.0). The combined effects would yield an intermediate exponent that could explain the observed exponent 0.75.

Table 7.4. Metabolic rate (P, in watts) relative to body mass (M_b, in kg) for major groups of higher vertebrates, expressed by allometric equations.

Animals	Allometric equations
Marsupial mammals	$P = 2.36 \times M_b^{0.737}$
Eutherian mammals	$P = 3.34 \times M_b^{0.75}$
Nonpasserine birds	$P = 3.79 \times M_b^{0.723}$
Passerine birds	$P = 6.25 \times M_b^{0.724}$

Is this theoretically plausible? If mammals had evolved in the absence of gravity, they presumably could be geometrically similar. Because transport of material in the body takes place across surfaces, gases in the lung, food in the intestine, materials crossing cell membranes, and so on, we can suggest that the fundamental metabolic rate should follow a surface law. Gravity, however, imposes a metabolic cost that must be added, and the total metabolic rate should therefore be the sum of the surface-related metabolism and the metabolic cost of living in terrestrial gravity (Economos, 1979). The suggestion is intriguing, although difficult to evaluate, because our present understanding of the complexities of metabolic rate is insufficient. However, it has been suggested that the hypothesis could be experimentally evaluated in the weightlessness of space (Pace and Smith, 1981).

Metabolic similarities

The metabolic rate characteristic of each major group of warm-blooded vertebrates can be represented by the logarithmic equation for metabolic rate versus body mass (Table 7.4). In these equations, all the exponents are nearly the same: between 0.72 and 0.75. Let us consider them as identical; the numerical coefficient then directly expresses the relative magnitude of the metabolic rate at any body size. The comparisons summarize what we have already discussed: that marsupial mammals generally have lower metabolic rates than eutherian mammals and that passerine birds, on the whole, have metabolic rates nearly twice as high as other birds of the same body size.

We saw that there are characteristic differences between these groups in regard to their normal body temperatures, the marsupials having the lowest and the passerine birds the highest body temperatures. We also

Table 7.5. Resting energy metabolic rates of higher vertebrates, recalculated to a uniform body temperature of 38°C (body mass, M_b, in kg, metabolic rate in watts) (from Dawson and Hulbert, 1970).

	Reptiles (lizard)	Mammals			Birds	
		Monotreme	Marsupial	Eutherian	Nonpasserine	Passerine
Body temperature (°C)	30	30	35.5	38	39.5	40.5
Resting metabolic rate (W)	$0.33\,M_b^{0.62}$	$1.65\,M_b^{0.75a}$	$2.36\,M_b^{0.737}$	$3.35\,M_b^{0.75}$	$3.80\,M_b^{0.723}$	$6.26\,M_b^{0.724}$
Q_{10}	3.3	2.1	2.5		2.5	2.5
Metabolic rate, recalculated to 38°C	$0.86\,M_b^{0.62}$	$2.99\,M_b^{0.75}$	$2.97\,M_b^{0.737}$	$3.35\,M_b^{0.75}$	$3.31\,M_b^{0.723}$	$4.98\,M_b^{0.724}$

[a] Assumed exponent = 0.75.

know that a change in the temperature of an organism changes its metabolic rate. Now assume that we change the body temperatures of these groups so that they are all at the same level. What would their metabolic rates be if all these animals were assigned a body temperature of 38°C? This paper experiment would permit us to compare at the same temperature the resting metabolic rates of the various higher vertebrates. The result is summarized in Table 7.5, which includes a similar calculation for lizards.

The recalculated or "temperature-corrected" coefficients show that, at 38°C, monotremes, marsupials, eutherian mammals, and nonpasserine birds all would have coefficients of similar magnitudes, because the lowest (2.97) and the highest (3.35) would differ from the mean by no more than 5%. The level for the passerine birds, on the other hand, would be 60% higher than the mean for the other four groups.

Lizards, however, even if their body temperatures were as high as those of mammals, would have metabolic rates at only one-quarter of the level for warm-blooded vertebrates. This suggestion is realistic, for many lizards easily tolerate being heated to 38°C, whereas monotremes, for example, do not. Thus, we can say that the metabolic rates of the so-called cold-blooded vertebrates, in this case reptiles, are at a lower level than those in warm-blooded vertebrates. This is not only because cold-blooded vertebrates have lower body temperatures but also because they have other metabolic characteristics different from those of warm-blooded animals. The latter, in addition to their well-insulated body surface, have resting metabolic rates at a level several times as high.

In this chapter we have considered overall energy metabolism, and it is reasonable to proceed to the next chapter to examine metabolism of the individual organs and tissues that make up the whole animal.

8

Organ size and tissue metabolism

The regular decrease in specific metabolic rate with increasing body size must somehow be reflected in the metabolic rates of the various organs that make up the whole animal. To extend this reasoning, the observed differences should also be reflected in the metabolic rates of the cells that make up these organs. We could therefore ask if it is fruitful to study the scaling problem from the viewpoint of cell metabolism.

Tissue metabolism and cell size

The peculiar situation is that large and small animals have cells that are roughly of the same size, within an order of magnitude of 10 μm (Teissier, 1939). For example, a microscopist would be hard put to recognize differences between microscopic sections of a horse muscle and a mouse muscle, except that the mitochondrial density is higher in the muscle from the smaller animal.

Because cell size in various animals is much the same, independent of body size, a large organism is not made up of larger cells, but of a larger number of cells of roughly the same size. It could therefore be expected that, as more cells of the same size are added to make up a larger organism, the metabolic rate should increase in proportion to the increased number of cells. As we have seen, this is not so, and we must therefore explore alternatives.

Could the observed decline in specific metabolic rate with increasing body size be due to relative decreases in the size of those organs that are metabolically most active? Various organs have tremendously different metabolic rates, the most active organs (kidney, heart, brain) having rates some 100 times as high as less active organs (e.g., skin), and bone and adipose tissue may have even lower metabolic rates. The metabolic

Table 8.1. Metabolic rates of the major organs in situ of a man (body mass = 65 kg, total metabolic rate 90.6 W) (adapted from Aschoff et al., 1971).

	Organ size		Organ metabolism	
	(kg)	(% of body mass)	(W)	(% of total)
Heart	0.3 ⎫	0.4 ⎫	9.7 ⎫	10.7 ⎫
Kidney	0.3 ⎪	0.5 ⎪	7.0 ⎪	7.7 ⎪
Brain	1.3 ⎬ 5	2.0 ⎬ 7.7	14.5 ⎬ 65.6	16.0 ⎬ 72.4
Lungs	0.6 ⎪	0.9 ⎪	4.0 ⎪	4.4 ⎪
Splanchnic organs	2.5 ⎭	3.9 ⎭	30.4 ⎭	33.6 ⎭
Muscle	27.0 ⎫	41.5 ⎫	14.2 ⎫	15.7 ⎫
Skin	5.0 ⎬ 60	7.8 ⎬ 92.3	1.7 ⎬ 24.0	1.9 ⎬ 27.6
Other	28.0 ⎭	43.0 ⎭	9.1 ⎭	10.0 ⎭
Total	65.0	100.0	90.6	100.0

rates of various organs are not well known for any wide variety of animals, but what is known about humans will be helpful. Table 8.1 shows that nearly three-quarters of the total metabolic activity in a 65-kg man takes place in organs with a combined weight of no more than 5 kg. The relative size of these highly active organs decreases with increasing body size; in a mouse the liver is 6% of the body mass and in an elephant about 1.6%. Can this explain the lower specific metabolic rate in the large animal?

The sizes of the metabolically most active organs in mammals are listed in the form of allometric equations in Table 8.2. We can now see that some (kidney, brain, liver) indeed decrease in relative size with increasing body size, but others (heart, lungs, skeletal muscle) maintain their relative sizes unchanged (body mass exponent = 0). We can compare the exponents to the specific metabolic rate for the whole animal, which is related to body size with the exponent −0.25. The brain is the only organ that decreases with a similar exponent; other active organs, such as kidney and liver, also decrease, but not to the same extent, and heart, lung, and muscle mass maintain roughly the same proportions, irrespective of body size. We can therefore immediately see that the observed decrease in specific metabolic rate cannot be explained by the decreases in the relative sizes of the metabolically most active organs.

Can we find, in the metabolic activity of the individual tissues, any other clue to the metabolic rate for the entire organism? There are two main theories:

Table 8.2. Allometric equations for the relative size of the metabolically most important mammalian organs expressed in percent of body mass, M_b (in kg) (data from Stahl, 1965).

Organ	Allometric equation
Kidney	$0.73\,M_b^{-0.15}$
Brain	$1.0\,M_b^{-0.30}$
Liver	$3.33\,M_b^{-0.13}$
Heart	$0.58\,M_b^{-0.02}$
Lung	$1.13\,M_b^{-0.01}$
Total muscle mass[a]	$40\,M_b^{0.0}$

[a] Approximate figure based on data from Munro (1969) and Smith and Pace (1971, Table 4).

1 The metabolic rate in homologous tissues (liver, for example) is relatively constant, irrespective of body size, but this rate is restricted or depressed in the larger animal by some "central" control or other "organismic" factor resident in the intact organism.

2 The metabolic rates for various tissues vary with body size to the same extent that total metabolic rate varies; thus, their sum will account for the decline in total metabolic rate with increasing body size.

One could hope that this question could be answered by studying the metabolic rates (actually oxygen consumption) of tissue slices in vitro. Early results strongly favored the first hypothesis. Terroine and Roche (1925), who studied brain, liver, kidney, and muscle from birds and mammals whose specific metabolic rates varied 10-fold, found that each kind of homologous tissue had the same respiration in vitro, irrespective of the animal it came from. The same conclusion was reached by Grafe and associates (1925a, 1925b), who studied nearly a dozen different tissues from mammals ranging in size from mice to cattle. Each homologous tissue had about the same respiration rate in all animals; the conclusion was that tissues from warm-blooded animals attain their energetic dependence on body size only within the living organism. In all probability, the strikingly uniform results obtained by Terroine and by Grafe resulted from inadequate techniques; the tissue respiration may have been diffusion-limited and thus may not have revealed any differences, for later studies with improved techniques gave different results.

A study published by Kleiber (1941) showed that metabolism of liver slices from rat, rabbit, and sheep indeed declined with increasing body size, the regression line having a slope of −0.24. This, of course, is strikingly similar to the specific metabolic rate for the whole organism (slope = −0.25). However, Kleiber omitted from his calculations data obtained from one cow and one horse; presumably he did not wish to use data from only one individual of a species, but if he had done so, the calculated slope would have been lower. Kleiber's results differed drastically from the earlier results and strongly supported the second hypothesis. Further studies, including more tissues as well as more animals, evidently were needed.

A major survey of tissue metabolism, using five major tissues from nine species of mammals, was published by Hans Krebs (1950). Krebs realized that, in addition to the importance of assuring adequate oxygen diffusion, the composition of the medium was important, especially in regard to metabolic substrates. Also, results differ when tissue slices and minced tissues are compared, some tissues giving higher oxygen consumption when minced, and others when sliced.

One question raised by Krebs was whether or not one could, by varying the conditions and substrates, determine a "standard" or "basal" rate of tissue metabolism, analogous to what one can determine for the whole animal. On the whole, the answer is negative, for there are too many variables to consider, and there is no firm support for any particular standard procedure that could apply to all tissues. For our purposes, this is relatively unimportant; our interest is in those results that pertain to the relationship between tissue metabolism and body size (Table 8.3).

We see that as the size of the animal increases from mouse to horse, the oxygen consumption of all tissues decreases. For the brain and the kidney, the decrease is to about one-half, and for liver, spleen, and lung, there is roughly a fourfold decrease. None of these is anywhere near the decrease in specific metabolic rate of the whole animal, which should decrease 14-fold over this size range.

The results quoted in Table 8.3 were obtained with a medium that contained several metabolic substrates (pyruvate, fumarate, glutamate, and glucose) but was low in calcium (Krebs medium II). A similar medium, but low in phosphate and containing calcium (Krebs medium III), gave a more pronounced size dependence for the respiration of the liver, but not for brain. Thus, changing the medium may change the body size dependence for one kind of tissue but not for another. Evidently, the search for a standard method for determining tissue metabolism is not easy.

Table 8.3. Rates of oxygen consumption for tissue slices from nine species of mammals (Krebs, 1950).

Species	Body mass (kg)	Relative rates of O_2 consumption[a]				
		Brain	Kidney	Liver	Spleen	Lung
Mouse	0.021	32.9	46.1	23.1	16.9	12.0
Rat	0.21	26.3	38.2	17.2	12.7	8.6
Guinea pig	0.51	27.3	31.8	13.0	11.6	8.5
Rabbit	1.05	28.2	34.5	11.6	14.2	8.0
Cat	2.75	26.9	22.7	13.2	8.4	3.9
Dog	15.9	21.2	27.0	11.7	6.6	4.9
Sheep	49	19.7	27.5	8.5	6.9	5.4
Cattle	420	17.2	23.5	8.2	4.4	4.3
Horse	725	15.7	21.5	5.4	4.2	4.4

[a] The original article did not report the units in which oxygen consumption was expressed.

The conclusions we can draw from Krebs's survey of tissue metabolism are, however, rather unequivocal. Although tissue metabolism determined in vitro does decline with increasing body size, the decline is nowhere near the decline in specific metabolic rate for the whole animal. This raises the question whether metabolism of isolated tissue slices has any bearing on metabolism of the same organ in the intact animal. An attempt to answer this question has used the following approach.

Summated tissue respiration

If the metabolic rate of tissue slices could give a correct figure for the metabolic rate of the organ from which the tissue was removed, it should be possible to add up all tissue metabolism determined for each kind of tissue and arrive at the total metabolic rate for the whole animal. This laborious procedure has been attempted only a few times, but the results are interesting and deserve our attention.

The first such study was carried out by a group of physiologists at Stanford University who had close contacts with Kleiber and at times collaborated with him on problems related to metabolism and body size (Field et al., 1939). This study examined young, postpubertal albino rats with an average body mass of 150 g. All distinct organs and tissues that could be dissected out were examined; careful precautions were taken to minimize loss of time, and the results were extrapolated to zero time (i.e., the moment of sacrifice of the animal). The overall results are listed in

Table 8.4. Summated oxygen consumption of a 150-g rat, calculated by adding the values for various organs as determined from tissue slices in vitro (Field et al., 1939).

Organ	Organ weight (g)	Rate of O_2 consumption [ml O_2 (g wet tissue hr)$^{-1}$]	Whole organ (ml O_2 hr^{-1})
Skeletal muscle	61.4	0.875	53.72
Diaphragm	1.0	1.800	1.80
Skin	27.8	0.416	11.55
Skeleton	10.0	0.153	1.53
Blood	9.7	0.025	0.24
Liver	9.2	2.010	16.48
Alimentary canal	8.0	1.010	8.08
Ligaments	7.4	0.070	0.52
Brain	2.3	1.840	4.23
Kidneys	1.4	4.120	5.76
Testes	1.2	1.030	1.24
Lungs	0.9	1.250	1.13
Heart	0.7	1.930	1.35
Spleen	0.4	1.330	0.53
Remainder	9.6	0.200	1.92
Total	150.0		110.08

Table 8.4, which shows metabolic rates for 14 different tissues. These add up to 110 ml of oxygen per hour, whereas the resting or standard metabolic rate of a 150-g rat is 167 ml O_2 hr^{-1}; that is, the summated tissue respiration accounts for 66% of the metabolism of the intact animal. Thus, the summated respiration for the isolated tissues of the rat is sufficiently close to that in situ that there is no reason to invoke any "organismic" factors that should control the level of tissue respiration in situ. On the other hand, the results do not contradict this theory. Discussions of these problems tend to be fruitless and a bit metaphysical (Bertalanffy and Pirozynski, 1951a, 1951b; Schmidt–Nielsen, 1951). Determinations on more animals of different body sizes might be helpful, for it is always easier to base a discussion on sufficient factual information.

An appropriate study was published in 1955 by Martin and Fuhrman, both previous members of the Stanford group. The fact that they extended the study to only two additional species, mouse and dog, should be viewed in light of the very laborious and time-consuming work involved

Table 8.5. Rates of tissue respiration for mouse, rat, and dog, expressed relative to tissue respiration for the same organ in the rat (Martin and Fuhrman, 1955).

	Mouse	Rat	Dog
Total metabolic rate [ml O_2 $(g\ hr)^{-1}$]	1.69	1.09	0.36
Relative oxygen consumption of tissues			
Brain	1.68	1.0	0.74
Bone	1.83	1.0	0.41
Digestive tract	1.79	1.0	0.67
Fat	2.15	1.0	1.30
Heart	0.95	1.0	0.55
Kidney	1.22	1.0	0.60
Liver	1.66	1.0	1.00
Lung	1.32	1.0	0.43
Muscle	1.44	1.0	0.65
Skin	1.15	1.0	0.41
Spleen	3.30	1.0	1.51
Testis	1.03	1.0	0.32

in complete dissection of an animal and determination of organ sizes and tissue metabolism. For the mouse (23 g), the summated tissue respiration accounted for 72% of the resting metabolic rate of the entire organism; for the dog (19.1 kg), the summated tissue respiration added up to 105% of the resting metabolic rate of the entire animal. Thus, there was again a reasonable correspondence between summated tissue respiration determined in vitro and the oxygen consumption of the intact animal.

The results for the three animals for which summated tissue respiration has been determined are listed in Table 8.5. Although there is a great deal of variability from one tissue to another, the respiration rates of mouse tissues are uniformly higher than for the rat, and the respiration rates of dog tissues are, for the most part, uniformly lower. However, there are a few exceptions, and it is difficult to say whether or not it is a coincidence that the figures add up to values as close to the oxygen consumption of the whole animal as indeed they do.

One thing should be noted. Because of the changing relative size of various organs with a change in body size, oxygen consumption for each tissue need not change with body size according to the same exponent as the metabolic rate of the whole animal. In any event, the results lend support to the concept that the respiration of tissue slices gives some approximation of the respiration of intact organs, although it certainly is not an accurate measure.

Table 8.6. Number of mitochondria in mammalian liver (Smith, 1956).

Number of mitochondria per gram liver $\propto M_b^{-0.1}$
Liver mass $\propto M_b^{0.82}$
Total number of mitochondria in liver $\propto M_b^{0.72}$
Number of liver mitochondria per gram of body mass $\propto M_b^{-0.28}$

Metabolic equipment of the tissues

We have now seen that oxygen consumption of various tissues does change with body size. Can this be correlated with important tissue components, such as mitochondria and metabolic enzymes?

An early survey of this problem was published by Drabkin (1950). He examined porphyrin chromoproteins in mammals ranging in size from rat to horse. He found that (1) the hemoglobin content of the organism is proportional to the body mass, an observation we shall return to later, and (2) the content of cytochrome c is proportional to body mass to the power 0.7 (for the animals from rat to cow). Thus, the higher concentration of cytochrome c in a small animal is in accord with its higher specific metabolic rate, the exponents for the regression equations for cytochrome c and specific metabolic rate being −0.3 and −0.25, respectively.

Cytochrome oxidase in rat, sheep, swine, and cattle was studied by Kunkel and associates (1956). The muscle content of cytochrome oxidase was related to body mass with the exponent −0.24. This is the same exponent as for the specific metabolic rate of the intact animal. Again, an important metabolic enzyme is present in higher concentrations in those animals that have the higher specific metabolic rates. Similar studies by Jansky (1961, 1963) showed the same trend; cytochrome oxidase activity was closely correlated to metabolic rate over a wide range of body size.

Another interesting study was concerned with the number of mitochondria in animals of various body size (Smith, 1956). This study included only four species (rat, rabbit, sheep, and cattle), but the results are convincing (Table 8.6). The density of mitochondria in the liver decreases with body size with an exponent of −0.1; thus, mitochondrial density decreases with increasing body size. However, the relative size of the liver is not constant; it decreases with increasing body size (line 2). The total number of mitochondria in the liver is found by multiplying lines 1 and 2, and we find that it is proportional to the body mass to the power 0.72 (line 3). Thus, the total number of liver mitochondria per gram body mass (line 4) decreases with increasing body size with the exponent -0.28, which is strikingly similar to the exponent for specific metabolic rate of the entire animal, -0.25.

This relationship between the mitochondrial equipment of the liver, one of the metabolically most important organs, and the total metabolic rate gives strong support to the concept that the metabolic equipment of tissues corresponds to the needs of the total organism. It does not lend support to the notion that tissue metabolism in the living animal is governed primarily by unknown "organismic" factors.

This conclusion is supported by a more recent study of muscle mitochondria. The mitochondrial densities in four different muscles from 13 species of mammals were always inversely related to body size. Scaling the total volume of mitochondria in the muscles to body size gave regression lines whose slopes closely paralleled the maximal rates of oxygen consumption (Mathieu et al., 1981).

It has been suggested that the oxidative capacity of mitochondria is directly related to the density of the cristae. In the above study, the packing of the cristae did not vary systematically with the aerobic capacity of the muscle fibers. Bartels, in contrast, reported that the density of cristae in the heart muscle of the smallest mammal, the Etruscan shrew (2.5 g body mass) is several times higher than in larger mammals (although the concentration of mitochondria is not significantly different) (Bartels, 1980). Evidently, to clarify these discrepancies we need additional quantitative studies of the mitochondria and the metabolic machinery.

One further study should be mentioned. The concentrations of three enzymes that function in oxidative metabolism in mammalian muscle, studied over a body size range of nearly six orders of magnitude, showed a clear negative correlation with body size (Emmett and Hochachka, 1981). However, the slopes of the regressions were less than those for specific metabolic rates. This may indicate that the slope for aerobic activity could be greater in large than in small mammals, but this suggestion does not agree with observations on animals running on treadmills when they reach maximal rates of oxygen uptake (Taylor et al., 1981).

To summarize the sparse information we have about the metabolic machinery that in the end is responsible for the use of oxygen, we can conclude that there is a close correlation with the metabolic rate of the whole animal. However, the information we now have is totally inadequate, and further studies are much needed.

9

How the lungs supply enough oxygen

The metabolic rate of an animal is maintained through the steady consumption of fuel and oxygen. From the viewpoint of scaling, much attention has been given to the supply of oxygen, and relatively little to the supply of fuel. In this and the next two chapters we shall focus on the gas transport system, especially as it pertains to oxygen.

First we shall consider the effects of scale on gas-exchange organs: lungs and gills. These organs must have a size and diffusion capacity adequately scaled to the need for oxygen. In the next chapter, the role of blood in gas transport will be considered, and because information is available mostly for vertebrates, in particular for mammals, the discussion will be limited to these. Following that we shall examine the circulatory system, which consists of a pump (the heart) and conduits through which blood is pumped.

When we analyze the function of the respiratory system from the viewpoint of scale, we should keep in mind that most of the available information has been obtained for animals at rest. The normal animal, if not asleep, spends much of its time being active and moving about, and the supply system for oxygen must be scaled to meet these extra demands (except during brief periods of non-steady-state conditions, such as a burst of maximal activity).

The lungs of mammals

In preceding chapters we saw that small mammals have higher specific metabolic rates than large mammals. Do small mammals, in order to meet the higher demand for oxygen, have relatively larger lungs? The answer is a simple no.

Figure 9.1. Logarithmic plot of lung volume as a function of body size. From Tenney and Remmers (1963).

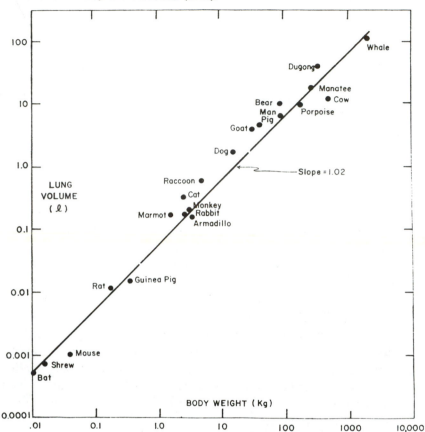

The lung volumes of various mammals are plotted on logarithmic coordinates in Figure 9.1. The points fall along a straight line, and the slope of the regression line is 1.02. In other words, the lung volume for small and large mammals makes up very nearly the same fraction of body volume (Tenney and Remmers, 1963).

A more recent compilation gave the following equation for the relationship of mammalian lung volume (V_ℓ, in ml) to body mass (M_b, in kg) (Stahl, 1967):

$$V_\ell = 53.5 \, M_b^{1.06 \pm 0.02}$$

This equation tells us the same: that lung volume relative to body mass is nearly constant, although there appears to be a slight trend toward an increase in relative lung volume with increasing body size.

Table 9.1. Respiratory variables in mammals, expressed as allometric equations (body mass, M_b, in kg) (data from Stahl, 1967).

Variable	Equation	Equation number
Tidal volume, ml	$V_t = 7.69\,M_b^{1.04}$	(1)
Vital capacity, ml	$VC = 56.7\,M_b^{1.03}$	(2)
Ventilation rate, ml min^{-1}	$\dot{V} = 379\,M_b^{0.80}$	(3)
Total compliance, ml (cm H$_2$O)$^{-1}$	$C = 1.56\,M_b^{1.04}$	(4)
Elastic work per minute (power), g cm min^{-1}	$\dot{W} = 962\,M_b^{0.78}$	(5)
Frequency of respiration, min^{-1}	$f_{resp} = 53.5\,M_b^{-0.26}$	(6)
Oxygen consumption rate, ml min^{-1}	$\dot{V}_{O_2} = 11.6\,M_b^{0.76}$	(7)

At the end of exhalation, a significant volume of lung air remains in the trachea and hence returns to the lungs on the next inhalation. This volume of "dead" air is known as the dead space. Its importance in mammalian respiration was evaluated by Tenney and Bartlett (1967). They examined animals ranging in size from shrews and bats to whales and found that the tracheal dead space is very nearly a constant fraction of the lung volume.

The conclusion is that the dead space has the same relative effect on lung ventilation in all mammals, and as a consequence the alveolar CO_2 concentration (P_{CO_2}) should be constant and independent of body size. This conclusion is, at least as a gross approximation, consistent with available information about lung and tissue gas concentrations.[1]

On the whole, the mammalian respiratory system is rather well known, and much of the available information can be used to derive allometric equations. Some such equations are listed in Table 9.1, and it is interesting to see what each reveals about mammalian respiration. However, this is not all, for the equations can be used for a number of further generalizations that are not immediately obvious; see, for example, Drorbaugh (1960) and Spells (1969). We shall examine Table 9.1 with this in mind.[2]

Equation (1) tells us that the tidal volume, the volume of a single inhalation, is very nearly proportional to the body size. For a 1-kg animal,

1 It has been suggested that very small mammals tend to have lower blood and alveolar P_{CO_2} than larger mammals (Lahiri, 1975). However, information on tissue P_{CO_2}, obtained from tissue air pockets, suggests that shrews, pocket mice, and bats in this regard fall within the usual mammalian range (Tenney and Morrison, 1967).

2 In Table 9.1 and in the following discussions the units will be those most commonly used in respiration physiology.

the tidal volume at rest will be about 8 ml, and for larger and smaller animals the proportion relative to body size will remain almost the same, for the exponent is not significantly different from 1.0. We saw before that the lung volume, V_ℓ, is also a constant fraction of the body volume. It therefore follows that each breath at rest renews the same fraction of the air in the lung. We shall return to this later.

The other equations can be analyzed similarly; each describes a respiratory variable as it applies to mammals in general. However, we can extract a great deal of additional information and insight by a combination of several of the equations. In some cases the result is a power law (or allometric prediction) for a physiological variable; in other cases it is possible to eliminate the body mass as a variable. In this way we can obtain a parameter that is independent of body size, a dimensionless number that gives a general characterization of mammalian respiration. Let us try some of the equations in Table 9.1.

By dividing equation (1) by equation (2), we obtain

$$\frac{V_t}{VC} = \frac{7.69\, M_b^{1.04}\ (\text{ml})}{56.7\, M_b^{1.03}\ (\text{ml})} = 0.136 = \frac{1}{7} \quad (\text{RME} = 0.01)$$

In these two equations, the mass-dependent variable has virtually the same exponents, and in the division they therefore cancel out, giving a residual mass exponent (RME) of 0.01, which is insignificant. The ratio is dimensionless as well as independent of body size. It tells that the normal tidal volume (V_t) for a mammal at rest is approximately 1/7 of its respiratory vital capacity (VC), whatever its body size.

Let us next examine the elastic properties of the respiratory system, as expressed in its total compliance, equation (4). Compliance is the change in volume of the respiratory system with the application of unit pressure (in traditional units, the number of milliliters the lung will expand when a pressure of 1 cm H_2O is applied). If we divide equation (4) by equation (2), we obtain what is known as specific compliance, that is, the change in volume undergone by each unit volume of lung with the application of unit pressure. The division is as follows:

$$\frac{C}{VC} = \frac{1.56\, M_b^{1.04}\ (\text{ml/cm}\,H_2O)}{56.7\, M_b^{1.03}\ (\text{ml})} = 0.028\ (\text{cm}\,H_2O)^{-1} \quad (\text{RME} = 0.01)$$

This equation tells us that a mammalian lung, for each milliliter of its volume, will expand by 0.028 ml when a pressure of 1 cm H_2O is applied. To paraphrase this statement, because the volume change per unit pressure change is a constant fraction of the lung volume, the conclusion is that all mammalian lungs have similar elastic properties.

Dividing equation (1) by equation (4) gives the following result:

$$\frac{V_t}{C} = \frac{7.69\,M_b^{1.04}\ (\text{ml})}{1.56\,M_b^{1.04}\ (\text{ml/cm H}_2\text{O})} = 4.93\ \text{cm H}_2\text{O}\quad (\text{RME} = 0)$$

In this case, the residual mass exponent happens to be exactly zero; the conclusion to be drawn from this operation is that the pressure required for the intake of one tidal volume is the same in all mammals.

Equation (5) divided by equation (7) gives information about the work of breathing expressed as a fraction of the total energy metabolism of the organism. In this case, the units used in the two equations are different, and they must be recalculated to compatible units. Energy as gram-centimeter per minute is recalculated to joules per second (watts), and 1 ml O_2 when metabolized corresponds to 20.1 joules. The calculation will therefore look as follows:

$$\frac{\dot{W}}{\dot{V}_{O_2}} = \frac{962\,M_b^{0.78}\ (\text{g-cm/min})}{11.6\,M_b^{0.76}\ (\text{ml O}_2/\text{min})} = \frac{962 \times 981 \times 10^{-7} \times (1/60)\ (\text{W})}{11.6 \times 20.1 \times (1/60)\ (\text{W})}$$

$$= 4.05 \times 10^{-4}\quad (\text{RME} = 0.02)$$

The ratio of these two equations is the ratio between the power required for breathing and metabolic power, in other words, a dimensionless ratio that in fact describes the efficiency of breathing. It says that the power required for breathing at rest is only 0.04% of total metabolic power. In evaluating this figure, we should recall that muscular efficiency is some 20%, and the minimum power required for breathing at rest will therefore be 0.2% of metabolic power. Other inefficiencies may also play a role, and the absolute magnitude may still be too low. However, the important conclusions are that the fraction of the metabolic power required for breathing at rest is a small fraction of the total metabolic power, and that for all mammals this fraction is a constant fraction of the metabolic power, independent of body size.

Equation (6) needs only a few words of comment. It will suffice to say that the frequency of respiration decreases with increasing body size with the same allometric exponent as specific metabolic rate, -0.26. This was discussed earlier as one of the consequences of McMahon's model of allometric scaling based on elastic similarity (Chapter 7).

As a final item, let us divide equation (7), the rate of oxygen consumption, by equation (3), the rate of ventilation of the lung. The result is

$$\frac{\dot{V}_{O_2}}{\dot{V}} = \frac{11.6\,M_b^{0.76}\ (\text{ml})}{379\,M_b^{0.80}\ (\text{ml})} = 0.031\quad (\text{RME} = -0.04)$$

In this case, the residual mass exponent is −0.04, which probably is insignificantly different from zero. The resulting number, 0.031, appears to be dimensionless, but it is not truly a dimensionless number, for the volume unit in one equation refers to oxygen and in the other to air. It says that for each unit volume of respired air, 0.031 volume of oxygen is metabolized (i.e., removed from the respiratory air). A more customary way of expressing this is to say that the oxygen removed in the lung is 3.1% of the respired air volume. The generalization we have arrived at is, within the accuracy of established relationships, that all mammals remove the same fraction of oxygen from the air they breathe.

These examples should make it clear what a powerful tool we have when we operate with allometric equations. They permit us to define similarities and derive general design criteria that are highly informative as well as simple.

It is, of course, necessary to keep in mind that an allometric analysis tends to suppress individual species differences. This is indeed desirable for a statement of overall design criteria, and when these are known, information derived in detailed studies on individual species can be compared to the generalized statement. Such comparisons will often pinpoint physiological specializations and may therefore be especially revealing and meaningful.

Bird lungs

The respiratory system in birds is radically different in design from the mammalian system. The lungs of mammals are essentially sacs with many subdivisions (alveoli) in which the air moves in and out. The bird lung, in contrast, consists of a large number of parallel tubes (parabronchi) through which air flows more or less continuously and in the same direction during both inspiration and expiration. The unidirectional throughflow of air in bird lungs thus differs drastically from the tidal flow in the mammalian lungs (Schmidt–Nielsen, 1975*b*). Whatever the relative merits of the two designs may be, they serve the same function, and because metabolic rates of birds and mammals of the same body size are similar, both designs serve gas exchange at the same rate, at least when the animal is at rest.

A remarkably comprehensive compilation of the scattered information on bird respiration was published by Lasiewski and Calder (1971), and another dealing specifically with tracheal dead space was published by Hinds and Calder (1971). Some of the allometric equations that these authors compiled are shown in Table 9.2, with similar equations for mammals listed for comparison.

Table 9.2. Allometric equations for respiratory variables for nonpasserine birds and for mammals (body mass, M_b, in kg) (data from Lasiewski and Calder, 1971; Stahl, 1967; Tenney and Bartlett, 1967).

Variable	Birds	Mammals	Equation number
Lung volume, ml[a]	$V_\ell = 29.6 M_b^{0.94}$	$53.5 M_b^{1.06}$	(8)
Tracheal volume, ml	$V_{tr} = 3.72 M_b^{1.09}$	$2.76 M_b^{1.05}$	(9)
Tidal volume, ml	$V_t = 13.2 M_b^{1.08}$	$7.69 M_b^{1.04}$	(10)
Respiratory frequency, min^{-1}	$f_{resp} = 17.2 M_b^{-0.31}$	$53.5 M_b^{-0.26}$	(11)
Ventilation rate, ml min^{-1}	$\dot{V} = 284 M_b^{0.77}$	$379 M_b^{0.80}$	(12)
Oxygen consumption rate, ml min^{-1}	$\dot{V}_{O_2} = 11.3 M_b^{0.72}$	$11.6 M_b^{0.76}$	(13)

[a] A more recent comparison (Maina and Settle, 1982) yielded equations for lung volume for birds ($V_\ell = 30.36 M_b^{1.048}$) and for mammals ($V_\ell = 41.92 M_b^{1.041}$) in which the mass exponents are nearly identical, but with a less striking difference in the proportionality coefficients than in this table.

Notice first the now familiar fact that birds and mammals have nearly identical rates of resting oxygen consumption [equation (13)]. In other equations we find striking differences between birds and mammals; specifically, birds have much smaller lungs [equation (8)], much lower respiratory frequencies [equation (11)], and correspondingly much higher tidal volumes [equation (10)].

On the whole, the body mass exponents in these equations seem to be similar for birds and mammals. Considering the heterogeneous material used for the calculations, the small differences in mass exponents are probably meaningless. The obvious differences are in the proportionality coefficients in equations (8), (10), and (11).

The comparison of birds and mammals can lead to a remarkable result. As we saw before, the metabolic rates for the two groups are similar [equation (13)]. Because the ventilation rate [equation (12)] in birds is lower, birds must remove a larger fraction of the oxygen in the respired air than mammals do. The oxygen removal can easily be calculated from equations (13) and (12):

$$\frac{\dot{V}_{O_2}}{\dot{V}} = \frac{11.3 M_b^{0.72}}{284 M_b^{0.77}} = 0.040 \quad (\text{RME} = -0.05)$$

This means that birds, in general, remove 40 ml of oxygen from each liter of respired air. We saw earlier that the corresponding figure for mammals is 31 ml of oxygen per liter of respiratory air. Thus, birds do indeed remove more oxygen from the respiratory air than mammals, and the exhaled P_{CO_2} must therefore also be correspondingly higher in birds. However, the peculiar situation is that the arterial P_{CO_2} in birds (which is about 30 mm Hg and size-independent) is lower than in mammals (40 mm Hg) (Calder and Schmidt–Nielsen, 1968).

How can this be explained? It seems to contradict the higher expired P_{CO_2} in birds. However, because of the peculiar arrangement of air flow in the bird lung, the exhaled air is not in diffusion equilibrium with the arterialized blood as it is in mammals. The arterialized blood may have higher oxygen and lower carbon dioxide partial pressures than correspond to equilibrium with exhaled air.

This apparent paradox is explained by the characteristic flow of the air and blood in the bird lung, which has been described as a cross-current flow (Scheid and Piiper, 1972). Thus, the allometric equations succinctly point out not only similarities but also characteristic differences between avian and mammalian respiration. In this case, the conclusions that were drawn from examination of the allometric equations pointed to differences that later were firmly established by direct experimentation.

A remarkably simple concept: symmorphosis

Consider the system that moves oxygen from the outside air to the oxygen-consuming mitochondria of the muscles. It consists of a series of steps: ventilation of the lung, diffusion of oxygen into the blood, circulation of blood, pumping by the heart, and diffusion from capillaries to mitochondria. Each component process must be capable of transporting oxygen at the maximal rate at which the muscles consume oxygen. On the other hand, there is no reason to maintain a higher transport capacity than required to satisfy the maximal need. For example, there is no need for the heart to be able to pump twice as much blood as will ever be required. The idea is simple: Each element in the chain should be structured to meet but not exceed the requirements of the others. This principle, that structures are designed to meet but not to exceed the maximal requirement, was called *symmorphosis* by Taylor and Weibel (1981).

This principle makes a great deal of sense. Building and maintaining structures in excess of what will ever be needed is expensive. For example, there is no need for more mitochondria in the muscles than needed for the maximal use of oxygen; there is no need for a larger heart than needed to pump the blood; the diffusing capacity of the lung need only be adequate for the amount of oxygen that the blood can carry away; and so on. This principle has already been implied in much of our discussion in previous chapters, although it has not been expressly stated.

According to this principle, we would expect that the maximal diffusing capacity of the lung should equal but not exceed the maximal rate at which the body can consume oxygen. (The fact that muscles, during a short burst or sprint, can exceed the maximal aerobic power is irrelevant; we are concerned with the match between the maximal steady-state use of oxygen and the ability of the lung to supply this oxygen.) It has long been suggested that the maximal rate of oxygen consumption of mammals is roughly 10 times the resting rate. This was confirmed by Taylor and associates (1981), who studied 14 species of wild mammals, ranging from 7-g pygmy mice to 217-kg elands. They obtained a regression line for maximal oxygen consumption relative to body size with a slope of 0.79 ± 0.05. When domestic animals were included in the calculations, the slope was similar, the mean maximal oxygen consumption being 10.3 times the resting oxygen consumption for any given body size.

Because the lungs must be able to supply oxygen at the maximal rate that it is used, the diffusing capacity of the lung for oxygen, $D_{L_{O_2}}$, should be matched to it. Measurements of diffusing capacity for mammalian lungs over a wide range of body sizes have given very interesting information (Gehr et al., 1981). Stated simply, extremely careful morphometric measurements have shown that diffusing capacity is scaled relative to body size with a slope of 0.99 ± 0.03 (32 species ranging from less than 3 g to 700 kg), not with a slope of 0.79 as could be expected from the maximal use of oxygen.

The discrepancy is clear and seems to be a paradox. Why is the diffusing capacity of the lung (slope $= 0.99$) scaled out of proportion to the maximal use of oxygen (slope $= 0.79$)? If the smallest mammal has an oxygen diffusing capacity just sufficient to supply the maximal rate of oxygen used (and it must have at least this much), it seems that the largest mammal has a diffusing capacity 10 or 20 times greater than will ever be needed. This is a paradox that seems to contradict the principle of symmorphosis. It is extremely unlikely that the morphometric measurements are wrong and contain a systematic error that changes with body size. Furthermore, direct measurements of diffusing capacity in living animals have confirmed the morphometric measurements (O'Neil and Leith, 1980). The explanation for the paradox must be sought elsewhere.

One possible reason for the paradox may be related to the assumed oxygen partial pressure in the alveolar air at the surface of the membrane across which oxygen diffuses into the blood. Although mean oxygen partial pressures in alveolar air presumably are similar in all mammals, around 100 mm Hg, the partial pressure at the very surface of the alveolar membrane could conceivably be lower in the larger animals.

Oxygen reaches the alveolar surface through two steps: bulk-flow ventilation of the lung, followed by diffusion within the alveoli. This diffusion implies an oxygen gradient within the alveoli and a lower oxygen concentration at the membrane surface. This gradient is often called *stratification*. If the path for diffusion increases with increasing body size, the concentration of oxygen at the membrane surface will be lower, thus providing a lower driving force for diffusion across the membrane in the larger animal. The apparently disproportionate diffusing capacity of the lung may therefore be related to a greater importance of stratification in the large animal (Weibel et al., 1981). The other possibility is that large mammals simply have more diffusing capacity than needed and that the apparent paradox is no paradox at all (Heusner, 1983).

Cold-blooded vertebrates

The respiratory systems of cold-blooded vertebrates are less well known. Some information about the lungs of amphibians and reptiles as related to body size has been compiled by Tenney and Tenney (1970). One disadvantage in that study was that the number of species did not suffice for calculation of reasonably precise regression equations; another difficulty was that it may not be appropriate to pool data from animals as different as turtles, snakes, and lizards, although the small number of measurements available made it desirable to do so in order to obtain more data points. Nevertheless, the data suggest that lung volume in cold-blooded vertebrates is more or less proportional to body size.

A study of oxygen exchange in amphibians of a wide size range was carried out by Ultsch (1973), but in that study a new complication arose. Gas exchange in amphibians is not served by the lungs alone, because the general body surface, the skin, is important in gas exchange. A further complication is that the relative roles of lungs and skin are not the same for exchange of oxygen and carbon dioxide (Hutchison et al., 1968).

A number of careful studies of vascularization of the skin as well as the lung in amphibians have given a clear indication of a close relationship between the rate of oxygen consumption and the vascularization of the skin. In small specimens, which have the highest specific metabolic rates, vascularization of the skin is higher than in larger individuals (Szarski, 1964). This relationship holds both for salamanders and for frogs and toads. The plethodontid salamanders, which have no lungs at all, have the highest number of skin capillaries (Czopek, 1965). The available information shows a clear body-size relationship, but it is not suitable for an analysis in the form of allometric equations.

For fish, however, the situation is different. A number of excellent studies have been carried out on the gills of fish, and the scaling of important variables such as gill area and diffusion distances has been well worked out.

Fish gills

For most fish, the gills are the major, if not the only, respiratory organ. The gills must permit adequate uptake of oxygen, and it is obvious that the size of this organ must change with the size of the fish, more specifically with the rate of oxygen consumption.

The physical principles of gas exchange in the fish gills and their relationship to the structure of the gills have been carefully studied by Hughes (1966) and by many other investigators; for a review, see Hughes and Morgan (1973).

The most important variable for the functioning of the gills in gas exchange is the *diffusing capacity*. [The dimensions for diffusing capacity in gills usually are given as $cm^3 O_2 min^{-1} (mm Hg)^{-1}$, which refers to the amount of oxygen that diffuses per unit time per unit partial pressure difference for the entire gill.]

The diffusing capacity is a function of the total surface area of the gills more than of any other single variable. It is difficult to determine the diffusing capacity directly (for example, the measured diffusing capacity varies with the rate of water flow between the gill lamellae), and we shall therefore examine only the surface area of the gills. The surface area is relatively easy to determine, although if the procedure is carefully executed it is highly laborious, as most morphometric studies are.

It has been well established that the relative gill area is much greater in highly active fish, such as tuna and mackerel, than in very sluggish fish, such as puffer and toadfish (Gray, 1954). This is, of course, in full agreement with the much higher rate of oxygen consumption in the highly active fish. However, we are not particularly interested in comparing gill areas of active and sluggish fish, for this is not a problem of scaling.

We are, however, interested in the function of the gill in relation to the size of the body and the rate of oxygen consumption. To avoid the problem of comparing how fish differ in their normal activities, we could seek to relate the gill area and the rate of oxygen consumption for fish of various sizes within one species. This could indeed be very informative, for many fish, as they grow, retain their shape unchanged and thus remain isometric over a wide size range. It would be interesting if we could relate gill size to metabolic requirements for oxygen during the maximal steady-state rate of oxygen consumption during swimming. In this field, however, information is available only for very few species, and these are not the species for which quantitative data on gill morphology have been carefully collected. Could we instead compare resting oxygen consumption for the fish and for its respiratory organ in the way we did so successfully for mammals?

The problem is that it is difficult, if not impossible, to define a "normal" resting metabolic rate for fish. The metabolic rate of a fish depends on the temperature of the water and can change by one to two orders of magnitude with a change in temperature. The metabolic rate also varies with the oxygen content of the water, with the time of measurement relative to the circadian cycle in the metabolic rate, with the time after introduction of a fish into the experimental apparatus, with the degree of suppression of external stimuli, with the level of light, with the size of the experimental chamber, and with the presence or absence of other fish (the "group effect"; goldfish and trout, for example, are quieter in groups).

In addition to all these factors, there are long-term effects from factors such as the thermal history of the fish, the previous light cycle, and so forth. Fish physiologists are well aware of the difficulties, and instead of trying to measure a "true" metabolic rate for fish, they have carefully defined terms such as basal metabolism, standard metabolism, resting metabolism, minimal resting metabolic rate, routine metabolism, and so on. However, even when conditions are carefully defined, it is difficult to compare the results of different investigators and the results for the many different species of fish.

The two factors discussed (the difficulty in establishing a normal metabolic rate for any one species and the extreme species-to-species variation) make it virtually impossible to compare data in the way mammalian data have been used for broad generalizations. What we can use for our specific purposes, however, is the dependence of metabolic rate on body size within one species, as determined under otherwise identical conditions, preferably by a single investigator. The voluminous literature on metabolic rates for fish shows wide variation in the slopes of regression lines that relate oxygen consumption to body size on log-log scales.

Instead of giving a complete list, the results can be briefly summarized as follows: The slopes (the exponent b in the usual metabolic equation $\dot{V}_{O_2} = a\,M_b^b$) are extremely variable, ranging from 0.37 to 1.05. Values at the extremes of this range are unusual, however, and most of the values fall within the range 0.7 to 0.9, with a preponderance of values close to 0.8. It is perhaps reasonable to summarize the many results by saying that the rate of oxygen consumption for fish appears to be neither directly surface-related ($b = 0.67$) nor directly proportional to body mass ($b = 1.0$). For most fish, the relationship seems to be intermediate between these two possibilities, and values outside this range are rare [data compiled from Barlow (1961), Beamish and Mookherjii (1964), Beamish (1964), Brett (1964), Edwards et al. (1969), Job (1955), Pritchard et al. (1958), Ralph and Everson (1968), Saunders (1963), and Wohlschlag (1963)].

Gill area. Let us return to the question of the surface area of fish gills. A number of careful studies have been carried out on several species of fish, ranging from less than 1 g to 40 kg in size. The data are listed in Table 9.3 and are quite informative. For some of the species, the size range is more than two orders of magnitude, which makes the data more meaningful. Let us first examine the coefficient a, which is a direct expression of the relative surface area of the gills. For a fish weighing 1 kg, a represents the numerical value of the gill surface area in square meters. It is therefore immediately apparent that the surface area of a 1-kg tuna, a fast-swimming fish, is 10 or 15 times as great as that of a 1-kg roach or toadfish, both relatively sluggish and inactive fish. Other data, not included in Table 9.3, suggest that species-to-species differences may be as great as 26-fold, because a 1-kg fish may have a gill surface area as small as 0.07 m^2 (Muir and Hughes, 1969).

In contrast to the interspecies variations in gill surface area, we find a remarkable constancy in the size dependence within any one species. The slopes of the regression lines range between 0.8 and 0.9. This is far less variable than the rates of oxygen consumption for fish, but as we noted earlier, the most common slopes for the metabolic regression lines were also in the range of 0.8 to 0.9.

Let us now compare the respiratory surface area for fish to that for mammals (Figure 9.2). The gill area for a tuna is, as we saw earlier, substantially higher than for the other fish that have been measured, and it approaches the lung area for mammals. Tuna are strong and fast-

Table 9.3. Respiratory surface areas for fish: gill surface area (A, m^2) related to body mass (M_b, kg) by the equation $A = aM_b^b$ (data for mammals included for comparison).

Species	Size range (kg)	a	b	References
Roach	—	0.129	0.9	Muir and Hughes (1969)
Toadfish	0.015–0.8	0.131	0.79	Hughes and Gray (1972)
Black bass	0.001–0.9	0.196	0.79	Price (1931)
19 species	0.071–6.4	0.399	0.82	Gray (1954); Ursin (1967), quoted by Muir and Hughes (1969)
Tuna (yellowfin and bluefin)	4–40	1.33	0.87	Muir and Hughes (1969)
Tuna (skipjack)	1–6	1.85	0.85	Muir and Hughes (1969)
Mammals	0.025–25	3.31	0.98	Weibel (1972)

Figure 9.2. Total surface area of fish gills related to body size. The sluggish toadfish has the lowest gill area; the fast-swimming tuna has the highest. Tuna, in fact, approach the mammals in total respiratory surface area. Data from Gray (1954), Hughes and Morgan (1973), and others.

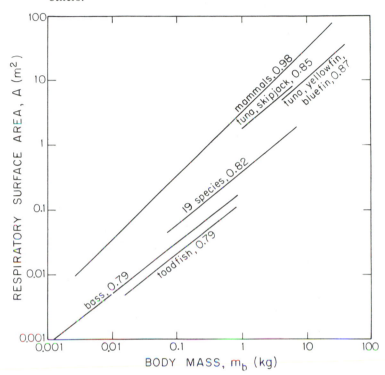

swimming fish, and perhaps most important, all the three species included in these measurements are so-called "warm-blooded" fish that regulate the temperature of their swimming muscles relatively independent of the water in which they swim (Carey and Teal, 1966, 1969; Graham, 1973).

We know that oxygen is present in water in amounts far smaller than in atmospheric air. Should we not expect, then, that a highly active, warm-blooded fish would need a much larger respiratory surface area than a mammal? A moment's thought will show that this should not be so. In well-aerated water, the oxygen is present in equilibrium with the partial pressure in atmospheric air, 0.2 atm. In the thin surface layer of fluid on the inner surface of the lung, oxygen will also be in diffusion equilibrium with the lung air (in this case, somewhat less than 0.2 atm).

Thus, the diffusion across the gill epithelium on the one hand, and the alveolar epithelium on the other, is from a surface film of water, across the epithelial membrane, to the blood on the inside. If the thicknesses of the epithelia are similar, the diffusion gradients will be similar, and the total area needed should be roughly the same for respiration in air and in water. (This assumes effective renewal of water, as indeed takes place in the gill; it would not be possible if an alveolar-type lung were filled with liquid).

To summarize this information, we can conclude that the surface area of the fish gill is related to the requirements for oxygen supply and that the surface area is scaled to the body size with a relationship similar to that for oxygen consumption. The species-to-species variations among fish and the difficulties in establishing an acceptable normal metabolic rate for any given fish make it difficult at the present time to analyze the scaling of fish gills with greater precision. Nevertheless, because of its well-defined, although complex, geometry, the fish gill should provide a fruitful area for further analysis, as initiated by Hills and Hughes (1970).

What we have seen in this chapter can be summarized as follows. The capacity of the respiratory system in vertebrates to supply oxygen is scaled in accordance with metabolic needs. This is not surprising; what is more important is that the equations that describe the scaling can be used to reveal characteristic differences as well as similarities in the general design principles of vertebrate respiratory systems. From this we shall move on to the medium that transports oxygen within the body: the blood.

10

Blood and gas transport

For all vertebrates and many invertebrates, the blood plays a major role in gas transport. In vertebrate blood, oxygen is carried by hemoglobin, which is located within the red blood cells. Carbon dioxide, in contrast, is carried mainly as the bicarbonate ion, dissolved in the blood plasma. It is commonly agreed that the supply of oxygen is more critical than the elimination of carbon dioxide and that whenever the oxygen supply is adequate, there is no difficulty in eliminating carbon dioxide at the rate at which it is formed.

The main parameters that are of interest in connection with oxygen transport and scaling are (1) the concentration of hemoglobin, which determines how much oxygen can be carried by one unit volume of blood, (2) the total volume of blood in the body and thus the total amount of hemoglobin in the blood, (3) the size of the red blood cell, and (4) the affinity of hemoglobin for oxygen, which is of interest both for the uptake of oxygen in the lung and for its delivery in the tissues.

Hemoglobin concentration

The hemoglobin concentration and oxygen capacity of blood have been surveyed by Larimer (1959) and Burke (1966) for a wide range of mammals. As could be expected, the oxygen-carrying capacity of the blood is strictly proportional to its hemoglobin concentration. The average hemoglobin concentration in 18 mammals, ranging in size from a small bat to the horse, was 128.7 g hemoglobin per liter blood. The corresponding mean oxygen capacity of the blood was 175.0 ml of oxygen per liter blood, with no distinct trend related to body size.

There is a great deal of additional information available in the literature. The overall picture is uniform: The hemoglobin concentration in

Figure 10.1. The viscosity of mammalian blood, relative to the viscosity of water (= 1.0), increases rapidly with increasing hemoglobin concentration. The optimal hemoglobin concentration for oxygen transport, relative to the work load on the heart, is at point A.

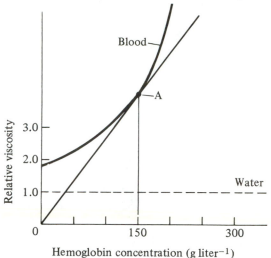

the blood in nearly all mammals seems to approach but rarely exceed about 150 g hemoglobin per liter blood, and this value seems to be independent of body size. Thus, the small mammal, in spite of its high specific metabolic rate and high relative demand for oxygen, has blood with the same oxygen-carrying capacity as large mammals.

What is the reason for this upper limit (which, however, in a few cases is exceeded)? Normal blood in mammals consists of some 45% red blood cells and 55% blood plasma, and the fraction of red cells is only rarely higher (Altman and Dittmer, 1961; Dunaway and Lewis, 1965). If humans become acclimatized to high altitude, the fraction of red blood cells, and thus the oxygen-carrying capacity of the blood, is increased in response to the lower oxygen pressure. An increased hemoglobin concentration in the blood carries with it a certain liability, for it increases the viscosity of the blood and puts an additional burden on the heart.

When the red cell fraction in blood is increased, the viscosity of the blood increases more and more rapidly, which makes it impossible to increase indefinitely the red cell fraction (Figure 10.1). Up to a certain point it pays to increase the hemoglobin concentration, because the viscosity, and therefore the work load on the heart, increases only moderately. Beyond this point, when the viscosity rapidly rises, the work load on the heart increases out of proportion to the increased oxygen-carrying capacity. The optimal point, at which the maximum amount of oxygen is carried per unit work by the heart, is given by the tangent to the viscosity curve drawn from the origin. The corresponding hemoglobin concentration is approximately 150 g per liter blood. This is also a common hemoglobin concentration in mammalian blood, which thus transports the maximum amount of oxygen with the smallest possible work load on the heart.

We saw that in altitude acclimatization the hemoglobin concentration is increased beyond this point; apparently the special circumstances lead to a transgression beyond the optimal hemoglobin concentration. It perhaps is no coincidence that the very smallest mammals, the shrews, have blood with a higher hemoglobin concentration than other mammals, about 170 g of hemoglobin per liter blood (Ulrich and Bartels, 1963). Small bats have even higher hemoglobin concentrations. Values up to 244 g hemoglobin per liter blood have been reported (Jürgens et al., 1981).

Our conclusion, then, is that the hemoglobin concentration in mammalian blood is scale-independent, but there may be special physiological demands that lead some animals to surpass what appears to be the optimal concentration.

Blood volume

The total volume of blood in most mammals is between 60 and 70 cm^3 blood per kilogram body mass; thus, relative blood volume is a size-independent parameter. The equation for blood volume (V_b, in ml) relative to body mass (M_b, in kg) calculated by Stahl (1967) is $V_b = 65.6 M_b^{1.02}$. The mass exponent is not significantly different from 1.0; that is, in general, the blood in mammals makes up a constant fraction of the body mass.

We should note that there are exceptions to this general rule, and these exceptions are strikingly meaningful. In particular, some diving mammals have much larger relative blood volumes; evidently this is because the blood provides the most important storage reservoir for the oxygen that is supplied to the brain during diving.

Because, in terrestrial mammals, blood volume is a constant fraction of the body, and the hemoglobin concentration of the blood is body-size-independent, the total amount of hemoglobin present in the body (the product of blood volume and hemoglobin concentration) will be a constant fraction of the body mass (again, with diving mammals as a major exception).

Table 10.1. Diameters of red blood cells
from various mammals; more than 100
additional mammalian species have red
blood cells within the range given here
(from Altman and Dittmer, 1961).

Species	RBC diameter (μm)
Shrew	7.5
Mouse	6.6
Rat	6.8
Dog	7.1
Sheep	4.8
Human	7.5
Cattle	5.9
Horse	5.5
Elephant	9.2
Humpback whale	8.2

Red cell size

The red blood cells of mammals are round, biconcave discs.
They are easily deformed and can pass through capillaries that have
smaller diameters than the diameter of the red cell, but the red cell's
diameter nevertheless gives a fair approximation to the necessary diameter for the blood capillaries. If we compare the size of red cells from
various mammals, we find the perhaps surprising fact that their diameters seem to be rather uniform and independent of body size (Table
10.1).

More extensive data on red cell size are available (Altman and
Dittmer, 1961). In a list of red cells from 115 species of mammals, most
of the red cells (over 100) were in the range of 5 to 8 μm, without any correlation with body size. The smallest mammal, the shrew, and one of the
largest, the humpback whale, have red cells of practically identical sizes
(7.5 and 8.2 μm, respectively). Not a single mammal has a red cell diameter over 10 μm, and only a few (sheep, goat, deer) have red cells less
than 5 μm.

The conclusions to be drawn from this information are two: that the
capillaries of the smallest and the largest mammals should have virtually
the same diameter, and that the high specific metabolic rates for the
smallest mammals do not seem to require any special adaptation in red
cell size (i.e., diffusion distance within the red cell is not scaled to metabolic rate).

Other vertebrates have oval, rather than circular, disc-shaped red cells. In birds, the shorter diameter of the oval red cell is similar to the diameter of mammalian red cells, 6 to 7 μm, and the longer diameter is about twice this. Reptilian red cells are also oval, and somewhat larger than bird red cells, with the shorter and longer diameters some 12 and 20 μm. Fish red cells are roughly in the same range, but amphibian red cells, by comparison, are of giant size. Frogs and toads have red cells that are 20 to 25 μm long, whereas salamanders have cells around 35 μm, with the maximum in some species exceeding 50 μm (Szarski and Czopek, 1966). At this time there is no clear understanding of what functional significance there may be in the very large amphibian red cells as compared with other vertebrate red cells, and we shall not discuss it further here.

Oxygen uptake and delivery

The reversible binding of oxygen to hemoglobin is of significance for two processes: (1) uptake of oxygen at the respiratory organ and (2) release of oxygen in the metabolizing tissues. Uptake of oxygen at the respiratory surface is favored by a high affinity of hemoglobin for oxygen; release of oxygen in the tissues, on the other hand, is favored by a low affinity, so that oxygen is readily released from its binding to hemoglobin. Thus, the affinity of hemoglobin for oxygen must be some kind of a compromise between these two requirements. Can we find any relationship between oxygen affinity and body size that is correlated with the high specific metabolic rates of small animals?

Oxygen affinity of hemoglobin. The customary and indeed very convenient way of describing the binding between hemoglobin and oxygen is the so-called oxygen dissociation curve. This curve shows, for any given sample of blood, the degree of saturation of the hemoglobin with oxygen at any given oxygen concentration. A number of typical oxygen dissociation curves for mammalian blood are shown in Figure 10.2.

The shapes of these curves are much the same, but their location in the usual coordinate system differs somewhat. If the dissociation curve is located to the left, the hemoglobin binds more oxygen at any given oxygen pressure. In other words, a curve located to the left indicates a higher affinity for oxygen. If the curve is located to the right, less oxygen is taken up at any given oxygen pressure; that is, the oxygen affinity of the hemoglobin is low, and oxygen is more easily released. Thus, the location of the dissociation curve to the left or to the right indicates high or low oxygen affinity, respectively.

Figure 10.2. Oxygen dissociation curves for blood from mammals of various sizes. Small mammals have a lower oxygen affinity. This helps in the delivery of oxygen in the tissues to sustain the high metabolic rate of a small animal. From Schmidt–Nielsen (1972).

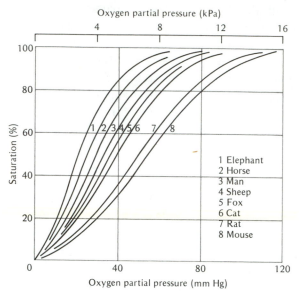

We saw in Chapter 9 that the ventilation of the lung is closely correlated to the need for oxygen, the consequence being that the oxygen partial pressures in the lung should be size-independent. In mammals, under normal circumstances, the partial pressure of oxygen in the alveolar air is very close to 100 mm Hg (at normal atmospheric pressure). A glance at the dissociation curves in Figure 10.2 shows that at a P_{O_2} of 100 mm Hg the blood of all mammals is virtually 100% saturated; that is, the hemoglobin has taken on a full load of oxygen and will bind no more, even if the oxygen pressure becomes higher.

We can conclude that, although mammalian bloods differ in oxygen affinity, this has no appreciable effect on how much oxygen can be taken up at the normal alveolar oxygen partial pressure.

It seems, then, that the range of oxygen affinities of mammalian blood is of little significance for the uptake of oxygen in the lung. This does not hold true, however, for animals that normally do not have a P_{O_2} of 100 mm Hg in their alveolar air. Thus, animals that normally live at high altitudes, such as the South American llama, have blood with much higher oxygen affinity than is usual for mammals; that is, the oxygen dissociation curve is located well to the left of the usual mammalian range (Hall et al., 1936). Thus, the high oxygen affinity of the llama's blood favors the uptake of oxygen at the low atmospheric pressures of high altitude. This high oxygen affinity and its effect on the degree of oxygen saturation in the arterial blood of the llama have been amply confirmed (e.g., Banchero et al., 1971). The shift to the left is characteristic of a few animals that normally live at high altitude; when lowland animals, including humans, become acclimatized to high altitude, their dissociation curves remain essentially unchanged or may in fact shift slightly in the opposite direction (Lenfant and Sullivan, 1971).

Some burrowing rodents have blood with exceptionally high oxygen affinity (dissociation curve located to the left). Among these, the prairie dog (a rodent) has especially high oxygen affinity, in fact, as high as that of the llama. Prairie dogs dig burrows that may be up to 5 m deep (Sheets et al., 1971), and presumably these animals are at times exposed to low oxygen pressures in their poorly ventilated burrows, with the high oxygen affinity of the blood facilitating uptake of oxygen. Indeed, the prairie dog can survive at a lower partial pressure of oxygen than any of the other 17 rodents studied (Hall, 1966).

In conclusion, for the large majority of mammals, the location of the oxygen dissociation curve of the blood (the oxygen affinity) is of little significance for the uptake of oxygen in the lungs at normal oxygen pressures. However, exceptions are found involving special environmental circumstances, such as altitude or burrowing habits, in which case oxygen uptake is favored by blood that has a higher oxygen affinity than usual.

Oxygen unloading in the tissues. The series of dissociation curves in Figure 10.2 suggest a relationship between oxygen affinity and body size of mammals. Small animals have dissociation curves located to the right, which indicates a lower oxygen affinity and easier unloading of oxygen, whereas larger animals have curves located to the left and higher oxygen affinities.

We have seen that the location of the dissociation curve is probably unrelated to the uptake of oxygen in the lungs, except in some special cases (altitude and burrowing animals). Is there, then, a more meaningful relationship between the ease of unloading of oxygen and body size?

From this viewpoint, the location of the dissociation curves may make more sense. However, we need a convenient way to express the location of the dissociation curve and the oxygen affinity. The customary

Figure 10.3. Relationship between body size and blood oxygen affinity determined at the body temperature for each species. From Dhindsa et al. (1971).

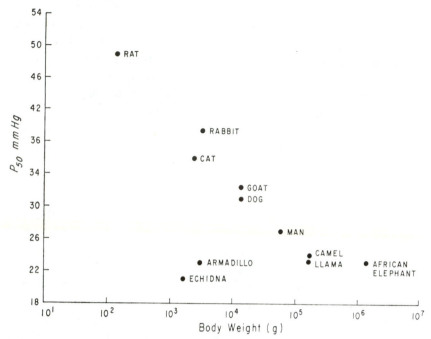

expression for this variable is the half-saturation pressure (P_{50}), which at times is referred to as the "unloading pressure" for oxygen. The P_{50} is not an exact unloading tension, for in some tissues the blood unloads less than half its oxygen, and in others, notably exercising muscles, a much greater fraction of the oxygen is unloaded. However, the half-saturation pressure can be used for a qualitative evaluation of oxygen affinity, for a curve located to the right has a higher unloading pressure than one to the left, whether one-quarter or three-quarters or more of the oxygen is given up.

How good is the evidence for a size-dependent oxygen affinity of the blood? Marsupials are like eutherian mammals; in seven species ranging from a 0.03-kg marsupial mouse to the 35-kg red kangaroo, the P_{50} ranged from 41.9 to 24.6 mm Hg (Bland and Holland, 1977). A clear size relationship was also reported for nine eutherians, as shown in Figure 10.3 (Dhindsa et al., 1971). However, two additional mammals, both burrowing animals, fell outside the range, having very high affinities.

Hall (1966) had earlier shown that many burrowing rodents have high oxygen affinities. There is, in fact, so much scatter and variability that Lahiri (1975) suggested that oxygen affinity is not well correlated with body size, and much of the variation seems to be explained by other factors (e.g., altitude or burrowing habits) (Prothero, 1980; Bartels, 1982).

A body-size relationship seems to exist in birds as well (Lutz et al., 1974), and perhaps also in frogs (Tazawa et al., 1979) and lizards (Pough, 1977). Lutz and associates studied seven species of birds, from sparrow (0.025 kg) to rhea (30 kg), a 1000-fold size range, and found P_{50} values from 41.3 to 20.7 mm Hg, with a highly significant relation to body size. Let us assume that in spite of the wide variability, there is a body-size relationship. What could be its functional meaning?

A higher oxygen unloading pressure in the small animal would be in accord with its high specific metabolic rate. Because the specific oxygen consumption of a shrew-sized mammal is some 30-fold greater than that of an elephant, the tissues of the shrew must be supplied with oxygen at a rate 30 times as high; that is, oxygen must diffuse from the capillaries to the metabolizing cells that much faster. This requires a steeper diffusion gradient, which can be achieved by (1) decreasing the diffusion distance and (2) increasing the diffusion head. The diffusion distance is a function of capillary density, as will be discussed later, and the diffusion head is the unloading pressure for oxygen (a function of the dissociation curve). The P_{50} for mouse blood and shrew blood is some two or three times higher than that for elephant blood, which is in accord with the foregoing. Why is the unloading tension (P_{50}) in the blood of the small animal not even higher? The answer is obvious: If the dissociation curve were located farther to the right, the blood would not become fully saturated in the lungs at the normal alveolar P_{O_2} of 100 mm Hg.

There is, however, another factor that also helps in the unloading of oxygen, the well-known Bohr effect. When CO_2 or other acid is added to the blood, the dissociation curve is shifted to the right, facilitating the unloading of oxygen. This effect increases the amount of oxygen unloaded in the tissues when CO_2 or other acid metabolites (such as lactic acid) diffuse into the blood in the tissue capillaries. It has been shown by Riggs (1960) that the Bohr effect is more pronounced for hemoglobin from small mammals than for hemoglobin from large mammals. In whole blood, the effect is less pronounced, but it is clearly greater in the smallest animals such as shrew and mouse (Bartels, 1964). Thus, the Bohr effect implements the already higher unloading pressure for oxygen in the tissues of the small animal.

The diffusion distance was mentioned earlier as an important variable in the diffusion gradient for oxygen from capillary to the consuming cell. The need for a high capillary density in small animals was understood by Krogh (1929), who estimated the numbers of capillaries in muscles from horse, dog, and guinea pig. The smallest animal had the highest capillary density (shorter diffusion distance), and this trend was confirmed by several later investigators. However, differences in technique make it difficult to compare the results of different authors, and we therefore decided to undertake a survey of capillary densities in several muscles from a large number of animals (Schmidt–Nielsen and Pennycuik, 1961). We found wide variations from muscle to muscle within one species, in part related to the proportions of red and white muscle fibers, and also to the common use of the muscle. For example, the exceptionally high capillary count in the chewing muscles of the sheep seem related to the fact that this ruminating animal spends much of its time moving its jaws. The relationship between capillary density and body size was less consistent than had previously been assumed. The very smallest animals did indeed have very high capillary counts, but larger ones, from the rabbit up, displayed no obvious size-related trend in capillary densities. It is possible that scaling considerations other than merely the supply of oxygen are more important than capillary distance, perhaps for entirely different reasons, such as the size of the muscle fiber and constraints on its size that secondarily influence capillary densities in ways we do not yet understand.

Fuel supply

So far, we have discussed metabolic rates only from the viewpoint of oxygen, but equally important is the supply of fuel or substrate to be oxidized. Unfortunately, this important aspect has often been neglected or totally disregarded. To maintain a high specific metabolic rate under steady-state conditions, the tissues must be supplied with fuel at a rate corresponding to the use of O_2. The very high sugar concentrations found in the blood of some insects can be interpreted in this light.

In insects, the transport of oxygen is through air-filled tubes, the tracheae, and is independent of blood and circulation, but fuel is supplied via the blood. Very active flying insects, such as honeybees, may have sugar concentrations in the blood as high as 5 or 10%, or two orders of magnitude higher than in most mammals (Wyatt, 1967), apparently to supply fuel for the highly active flight muscles.

The glucose concentration in mammalian blood seems to be scaled to body size, the higher concentrations being found in small mammals (Umminger, 1975). This would be analogous to the oxygen supply, in which the high rate of delivery to the tissues is achieved through a steep diffusion gradient, which in turn depends on high concentration differences as well as short diffusion distances. It would be interesting to see if this relationship is confirmed as more material is gathered, and also whether or not a similar relationship is found in birds.

Conclusions

To summarize the characteristics of blood that are important in oxygen transport, we can make the following statements:

1 The hemoglobin concentration in the blood of higher vertebrates is relatively constant and size-independent: about 15 g hemoglobin per 100 ml blood. This value is optimal in regard to reducing the work of the heart. Higher hemoglobin concentrations are associated with exceptional requirements, such as in diving animals, for which the demand for increased oxygen storage in the blood overrides the constraint on hemoglobin concentration.

2 The blood volume in mammals is between 6 and 7% of body volume and is independent of body size. Diving mammals, in which storage of oxygen in the blood is of importance, are again an exception; they have larger blood volumes.

3 The size of the red blood cell is, within each vertebrate group, unrelated to body size. Shrews and elephants have red blood cells of roughly the same size. There is no clear understanding of why each vertebrate group has a characteristic red cell size. The very large red cell of salamanders remains without rational explanation.

4 The oxygen affinity of the hemoglobin seems to vary with body size. This appears unrelated to the uptake of oxygen in the lung, except for some animals in which a higher oxygen affinity has adaptive significance (burrowing rodents and altitude animals, but not diving animals). On the other hand, the low oxygen affinity in mammals of small size facilitates the unloading of oxygen in the tissues. Oxygen delivery to the tissues of the smallest mammals is also aided by a higher capillary density, i.e., a shorter diffusion distance, and further by the effect of CO_2 in reducing the hemoglobin affinity for oxygen (the Bohr effect).

11

Heart and circulation

The circulatory system consists of a pump, the heart, and the attached plumbing of blood vessels. The dimensions of both pump and plumbing must be scaled to the demands, and it seems evident that in vertebrates the transport of gases is the most important consideration. If this demand is met, it appears that most or all other functional requirements on the circulatory system will be satisfied. This applies to the transport of nutrients, metabolic intermediates, excretory products, hormones, heat, and so on.

The circulatory system serves also in the transmission of force; blood is used as a hydraulic fluid to achieve, for example, ultrafiltration (in the kidney and in fluid exchange in the capillaries) and volume changes (e.g., erection of the penis). The needed force is supplied by the heart, as reflected in the blood pressure; we shall see that blood pressure appears to be a scale-independent physiological parameter.

The mammalian heart

The rate of oxygen consumption in mammals, relative to body size, decreases with increasing body size. It is therefore somewhat surprising to find that the relative size of the heart of small and large mammals is similar, that both mouse and elephant have hearts that are around 0.6% of their body mass. There is a great deal of information available on the size of the mammalian heart (e.g., Clark, 1927; Crile and Quiring, 1940; Grande and Taylor, 1965; Holt et al., 1968). The following equation for the mass of the heart relative to body mass (M_b in kg) represents the general picture (Prothero, 1979):

$$M_h = 0.0058 M_b^{0.98}$$

The exponent in this equation, 0.98, has 95% confidence limits of ±0.02, which means that for all practical purposes it is identical with unity. It states that in mammals in general the heart makes up the same fraction of the body mass, the numerical value being 0.58% of body mass, of course with a range of variations above and below the stated value.

All mammalian hearts are built on the same pattern, and for the moment we can assume that the stroke volume is proportional to the size of the heart itself. Thus, the decreased relative need for oxygen and blood flow in the large animal is not achieved through a relatively smaller heart or stroke volume, but through a decrease in heart rate.

The observed heartbeat frequency (f_h, in min^{-1}; M_b in kg) in mammals at rest gives the following equation (Stahl, 1967, based on data from Brody, 1945):

$$f_h = 241 M_b^{-0.25}$$

The exponent in this equation, −0.25, which has 95% confidence limits of ±0.02, is precisely the same as in the equation for specific metabolic rate. Consequently, the adjustment for the decrease in specific metabolic rate with increasing body size is entirely taken care of by the decrease in the rate of the pump, whereas heart size (and presumably stroke volume) remains an unchanged fraction of body size.

In this simple argument we disregarded the possibility of different oxygen capacities in the blood of small and large mammals, but we have already seen that the hemoglobin concentration and thus the oxygen capacity of mammalian blood are independent of body size.

Are these conclusions in accord with observations of cardiac output in mammals of different body size? We can examine the equation for cardiac output ($\dot{Q}_h$, in ml min^{-1}, and M_b in kg) given by Stahl (1967):

$$\dot{Q}_h = 187 M_b^{0.81}$$

The body-mass exponent, 0.81, has 95% confidence limits of ±0.01, and if the preceding arguments were entirely correct, we would expect the exponent to be 0.75 (as for the rate of oxygen consumption). However, the somewhat higher exponent may indicate that blood flow relative to oxygen consumption is somewhat higher in the large animal. It is difficult to establish whether or not these exponents are actually different; in addition to statistical uncertainty, there are other sources of variation, including variations in methods, in the physiological condition of the animals, and in particular in sampling or the selection of species (which

Table 11.1. Estimated arteriovenous oxygen differences for mammals of various body sizes, calculated from equations quoted in the text.

	Body mass (kg)				
	0.1	1	10	100	1000
O_2 consumption, ml min^{-1}	2.02	11.6	66.7	384	2210
Cardiac output, ml min^{-1}	28.96	187	1207	7795	50 332
Estimated arteriovenous O_2 difference (ml O_2/100 ml blood)	6.96	6.20	5.53	4.93	4.39

could be called biological noise). What does the difference in exponents mean in actual numbers? Table 11.1 compares calculated oxygen uptake with cardiac output calculated from the foregoing equation, for body sizes from 0.1 kg to 1000 kg. These figures are then used for an estimate of the arteriovenous oxygen differences in the blood of these animals.

The estimated arteriovenous oxygen difference comes out in the range of 4 to 7 ml oxygen per 100 ml blood, a perfectly reasonable range. Whether or not there is a real trend for smaller animals to utilize a higher fraction of the oxygen carried in the arterial blood remains an open question; the uncertainty of the exponents in these equations can only indicate the possibility of such a trend that may need investigation. However, there is a very real possibility that the trend points to a discontinuity that is related to physiological constraints on the function of the heart in the very smallest mammals.

Shrews, the smallest mammals. An adult shrew may weigh no more than a few grams. One of the smallest is the Etruscan shrew (*Suncus etruscus*), which weighs about 2.5 g (Weibel et al., 1971). Another is the masked shrew (*Sorex cinereus*), which weighs between 3 and 4 g. Calculation of the expected heartbeat frequency for a 3-g mammal suggests a heart rate of 1029 beats per minute for the animal at rest.

Before we proceed, it is worth noting that for shrews, the resting oxygen consumption, as far as it can be determined in such highly active and restless animals, is already more than three times higher than would be expected on the basis of body size (Table 11.2). Similarly high or even higher metabolic rates have been observed in other small shrews; a comprehensive compilation has been given by Vogel (1976). It is possible to

Table 11.2. Comparison of observed oxygen consumption and heart frequency for the masked shrew (body mass = 0.003 kg) (Morrison, Ryser, and Dawe, 1959); expected values calculated from standard regression equations according to Stahl (1967).

	Observed	Expected	Ratio observed : expected
O_2 consumption (liter O_2 g^{-1} hr^{-1})	9.0 (rest) 30 (max)	2.81	3.2
Heart size (% of body mass)	1.66	0.58	2.9
Heart frequency (min^{-1})	600 (rest) 1320 (max)	1029	0.6

suggest a rational explanation for this deviation from the "expected" metabolic rates for the very smallest mammals, a matter that will be discussed later under problems of heat and temperature regulation.

If the rate of oxygen consumption is three times the expected value, the heartbeat frequency should also be three times as high as the expected rate, or an impossible 3000 beats per minute. However, this assumes the usual scaling of stroke volume and heart size (0.58% of body mass). The masked shrew, however, has a disproportionately large heart, 1.7% of body mass, nearly three times the expected size (Table 11.2). It is now reasonable to ask if we are faced with a physiological constraint on how short the contraction time can be. A certain minimum time is needed for a heart to contract, expel blood, relax, and be filled again, ready for the next contraction.

The highest heartbeat frequency observed in shrews (and in small hummingbirds, which have approximately the same body size) is more than 1200 beats per minute. This means that less than 50 msec is available for the entire heart cycle. Our knowledge of conduction velocities and speeds of contraction for muscles suggests that these may well set ultimate constraints on heartbeat frequencies. If this is so, the disproportionately large size of the shrew heart is an expression of the impossibility of pushing the frequency high enough to keep up with the oxygen demand. An increase in heart size then becomes the only way out. Interestingly, the maximum heartbeat frequency in the masked shrew

($1320 \, \text{min}^{-1}$) is about twice the resting rate. In other mammals, the ratio between heartbeat frequencies at rest and in maximum exercise is somewhat higher, but not very different, approximately threefold.

We shall see that there is similar deviation from the expected heart size in the smallest hummingbirds; their hearts are disproportionately large as compared with other birds. Again, it is reasonable to suggest that there is a discontinuity in the scaling, because of constraints on the possible increase in heartbeat frequency.

The bird heart

Heart size for a wide variety of birds has been recorded by several authors (e.g., Parrot, 1894; Hartman, 1955). For our purposes, it is simplest to use the equations derived by Calder (1968) and Lasiewski and Calder (1971), who in their allometric analysis of respiratory variables in birds considered a wide range of available information. Their equations on heart size were based on the material recorded by Hartman (1955), who collected information on 1340 birds of 291 species. This material has the advantage of being collected by one investigator; variations introduced by using material collected by many different investigators are therefore avoided.

For the mass of the heart (M_h and M_b in kg) and heart frequency (f_h, in min^{-1}) for birds, the equations are

$$M_h = 0.0082 \, M_b^{0.91}$$

$$f_h = 155.8 \, M_b^{-0.23}$$

What can be read from these equations? First, consider cardiac output. Assume, as we did for mammals, that stroke volume is proportional to heart size, a very reasonable assumption in the absence of information to the contrary. By multiplying heart mass by heart frequency, we should obtain a number that is an expression for cardiac output. For a 1-kg bird, we can use the coefficient for heart size (0.0082), which multiplied by the coefficient for frequency (155.8) gives the number 1.278. A similar multiplication for mammals gives $0.0058 \times 241 = 1.398$. The similarity of these two numbers suggests that cardiac output in birds and mammals is similar, a similarity that could be expected, because the metabolic rates for birds and mammals of the same size are nearly identical, as discussed in Chapter 6.

Another way of stating this relationship is to say that the larger heart size for birds is offset by a corresponding lower frequency, at least at rest. What the situation is in exercise (for example, during flight) is not well known. It is often assumed that the heart frequency in mammals may increase up to threefold during maximal exercise. The few observations that exist for birds suggest a similar increase of up to threefold or a bit more. The information reviewed by Berger and Hart (1974) suggests that heart rates in flight are 2.4 times those at rest in small birds (10–20 g) and three times resting rates in large birds (500–1000 g). Higher frequencies were recorded in pigeons, with a mean rate during short flights of 3.5 times the resting rate (Hart and Roy, 1966). Much higher rates have been recorded in pigeons (115 at rest, 670 in flight), that is, a ratio of 5.8 (Butler et al., 1977), and in barnacle geese a ratio as high as 7.2 (Butler and Woakes, 1980).

In the absence of better information, it seems reasonable to assume, tentatively, that there may be constraints on heart rates in birds that differ from those for mammals. The lack of adequate information emphasizes how little we know about animals during activity, particularly during maximum performance, an area of rapidly expanding research activity. It certainly is an interesting area of research.

It is also worthwhile to examine the exponents in the foregoing equations. The slope of the regression line for heartbeat frequency in birds is -0.23, similar to those for the regression lines for specific oxygen consumption and also the regression lines for frequency and oxygen consumption in mammals. One would then perhaps also expect that heart size should have the same exponent as in mammals, close to 1.0. It appears, however, that the recorded exponent, 0.91, which has 95% confidence limits of ±0.03, is significantly different from unity. This simply means that the smaller the bird, the relatively larger is the heart. This was confirmed by Grubb (1983), who recorded an exponent of 0.92 over a 1000-fold size range.

The increase in relative heart size at the lower extreme of body size is the same phenomenon that we observed in the smallest mammals, the shrews, but for the birds the increase in relative heart size seems to be more evenly distributed over the size range. However, the hummingbirds are again, as were the shrews, distinctly above the uniform regression lines for birds in general.

The heart sizes recorded by Hartman (1955) for two species of hummingbirds are given in Table 11.3 and compared with the expected heart size for a bird of the same body mass, calculated from the equation given earlier. Thus, the heart of hummingbirds is substantially larger than suggested by the equation, which already accounts for a relative increase in heart size in small birds. Again, one is tempted to conclude that the

Table 11.3. Heart sizes for two species of hummingbirds compared with "expected" values calculated from the equation for bird hearts.

	Body mass (g)	Heart size (% of body mass)		Ratio observed : expected
		Observed	Expected	
Campylopterus hemileucurus	11.9	1.95	1.22	1.6
Selasphorus scintilla	2.23	2.40	1.42	1.7

exceptional size of the heart in these very small birds is related to constraints on maximum heartbeat frequencies.

The maximum heartbeat frequencies recorded for hummingbirds are similar to those for shrews: 1260 beats per minute for a 3.3-g hummingbird (Lasiewski, 1964). We have no assurance that these rates are truly maximal or that 1200 or 1300 beats per minute is a limit that cannot be exceeded. However, if heartbeat frequency is a constraint, the high blood flow required by a flying hummingbird probably can be met only by increasing the stroke volume (i.e., the size of the heart as well as the oxygen capacity of the blood). Additional information is needed to clarify this point.

Marsupials

We saw in Chapter 6 that marsupials have consistently lower metabolic rates than eutherian mammals, but their body-size dependence is similar to that for mammals (identical slopes of the regression lines). What circulatory adjustments go with this difference?

Heartbeat frequencies (f_h, in min^{-1}, and M_b in kg) for marsupial mammals have been recorded for animals over a 1000-fold size range, from 0.019 to 19.8 kg (Kinnear and Brown, 1967). The regression equation for the 17 species studied is

$$f_h = 106 \, M_b^{-0.27}$$

The slope of the body-mass exponent (-0.27) is not significantly different from that for eutherian mammals or from that for the specific oxygen-consumption rate. The numerical coefficient, however, is less than one-half of that for eutherian mammals: 106 against 241. Thus, marsupials in general have heart rates less than one-half of those for eutherians of the same body size; this is in accord with a lower resting metabolic rate for marsupials as compared with eutherian mammals (Dawson and Hulbert, 1970; MacMillen and Nelson, 1969). During exercise, the heart rate is increased. Six species of marsupials ranging from 0.112 to 26.7 kg had maximum sustained heart rates that were between 1.9 and 2.8 times higher than the resting rates in the same species (Baudinette, 1978). This is perhaps less than commonly found in eutherians, but its significance is difficult to evaluate in the absence of data on cardiac stroke volume.

In summary, observations on heart size and heartbeat frequency in marsupials indicate that the lower cardiac outputs that go with their lower metabolic rates are due to low heartbeat frequencies, whereas the size of the heart remains a constant fraction of body mass. Measurements of cardiac output would be interesting, because they could be used to calculate stroke volume. Another aspect that deserves study is the arteriovenous oxygen difference, which could readily differ from the eutherian norm.

Cold-blooded vertebrates

The hearts of reptiles and amphibians are somewhat smaller than mammalian hearts, on the average perhaps some 0.4 to 0.5% of body mass (Altman and Dittmer, 1971). I am not aware of any careful analysis over a wide range of body sizes, but I see no reason for a systematic deviation from the pattern in higher vertebrates, in which relative heart size is independent of body size.

The rates of oxygen consumption for reptiles and amphibians are an order of magnitude or more lower than in mammals, and they are also highly dependent on temperature. Nevertheless, because their hearts are not drastically smaller than those of mammals, the lower demands on cardiac output will be adjusted through lower heart frequencies.

However, what is a "normal" heart frequency for a cold-blooded vertebrate? For mammals and birds, the heart rate at rest is a rather constant value; for a cold-blooded animal, we can imagine that the heart rate could be measured at some temperature "normal" for this animal. However, what is a normal temperature for one animal might be very different from that for another. This makes it impossible to measure a normal rate of oxygen consumption for a range of cold-blooded animals.

For the fish heart, the extensive material collected by Crile and Quiring (1940) permits a better analysis of the relationship between heart and body size. Their data on heart size (M_h, in kg) for 34 species of fish ranging from 0.005 to 32 kg gave the equation

$$M_h = 0.0022 \, M_b^{1.026}$$

We see that the relative size of the heart in teleost fish is 0.22% of body mass and is virtually unchanged with body size. The typical 1-kg fish will have a heart of 2.2 g.

The information on heart size in elasmobranch fish is insufficient for a reliable comparison with the teleosts; in the four species recorded by Crile and Quiring (1940), the relative heart size ranged from 0.06 to 0.27% of body mass, with a mean of 0.15 ± 0.08 (SD).

What can we conclude from this meager information? First, within each major vertebrate group the relative size of the heart remains the same, independent of body size (except that the smallest mammals and birds have disproportionately larger hearts). Second, reptiles and amphibians have smaller hearts than mammals, and fish have even smaller hearts, about one-third the relative size of mammalian hearts.

Invertebrates

The relationship between heartbeat frequency and metabolic rate in mammals and birds has an interesting analogy in one group of invertebrates: spiders. For mammals and birds, we saw that heartbeat frequency varies inversely with body mass, the regression line having the same negative slope (-0.25) as that for specific metabolic rate.

A wide variety of spiders (ranging over more than two orders of magnitude in size, from about 30 to 10 000 mg) showed that heartbeat frequency (f_h, in min^{-1}) relative to body mass (M_b, in mg) gives the following equation (Carrel and Heathcote, 1976):

$$f_h = 423 \, M_b^{-0.409}$$

In the calculation of this equation, all kinds of spiders were included, except for a small group of primitive hunters and weavers. The authors then calculated the equation for the specific metabolic rate for spiders ($\dot{V}_{O_2}^*$, in ml O_2 g^{-1} hr^{-1}), using data from Anderson (1970, 1974), and reported the following equation (M_b, in mg):

$$\dot{V}_{O_2}^* = 947 \, M_b^{-0.408}$$

These mass exponents for heart rate and for specific oxygen consumption show a striking similarity; both are -0.41. This exponent is different from the corresponding exponent for mammals: -0.25. Otherwise the situation is as in mammals, in which the change in specific metabolic rate with body size is entirely adjusted through the heartbeat frequency, while the relative heart size remains unchanged. Presumably the relative heart size for spiders also remains constant, although information to verify this suggestion is not available.

The work of the heart

The work required to circulate the blood is provided by the heart. The large heart of an elephant obviously does more work than the small heart of a shrew, but exactly how does this variable scale with body size?

The work of a pump is usually considered to have three components: (1) pressure energy due to the internal pressure, (2) kinetic energy imparted to the moving fluid, and (3) potential energy due to gravity. For the heart, the gravitational potential energy can be disregarded, because the blood returns to the heart at the same level at which it is pumped out. The imparted kinetic energy is, at least at rest, a relatively small fraction of the total energy. We can therefore simplify the matter greatly and consider the work of the heart (W_h) as being the product of the mean arterial pressure (P) and the ejected volume of blood (V), or $W_h = P \times V$.

The work required for pumping blood through the pulmonary system of mammals is small compared with that required by the systemic circulation. The pressure in the pulmonary artery is on the order of 20% of the pressure in the aorta, and because the volume of blood moving in the two systems is the same, the work will be directly proportional to the mean arterial pressure in each system. That is, the systemic circulation requires about five times as much work as the pulmonary circulation.

We have already seen that cardiac output (minute volume) scales with body size as the oxygen consumption, or very close to it. All available evidence suggests that the blood pressure in mammals is a physiological invariable, the mean arterial blood pressure being about 100 mm Hg and independent of body size. As a consequence, the total work of the heart (mean arterial pressure times cardiac output) will be the same constant fraction of the total oxygen consumption, irrespective of the body size of the animal.

The next step is to compare birds and mammals. Again, as far as we know, blood pressure in birds is independent of body size, although it appears to be somewhat higher than in mammals. Grubb recorded the mean blood pressure in six bird species from pigeons to emus (0.4 to 38 kg) and reported an average of 133 mm Hg, compared with 97 mm Hg in mammals (Grubb, 1983). Because cardiac outputs in the two groups are similar, the work of the bird heart should be greater than that in mammals by a third. This is in accord with the larger size of the heart in birds as compared with mammals.

We should realize, of course, that these are very broad, sweeping generalizations that only indicate what the situation probably is. More information and more detailed studies are certainly needed in this area.

However, the conclusion that we reached by considering blood pressure can also be derived by analyzing familiar allometric equations. For the specific rate of oxygen consumption ($\dot{V}_{O_2}^*$) and for the heartbeat frequency (f_h), the equations for mammals are

$$\dot{V}_{O_2}^* = k_1 M_b^{-0.25}$$

$$f_h = k_2 M_b^{-0.25}$$

By dividing one equation by the other, we eliminate the body-mass variable and obtain the following general expression

$$\frac{\dot{V}_{O_2}^*}{f_h} = \frac{k_1}{k_2} = k'$$

The units in this equation are

$$\frac{\text{ml O}_2 \text{ g}^{-1} \text{ sec}^{-1}}{\text{sec}^{-1}} = \text{ml O}_2 \text{ g}^{-1}$$

and the residual mass exponent (RME) is zero. This result tells us that the amount of metabolic energy (ml O_2) consumed per gram body mass for one heartbeat is a constant, k', and is independent of body size.

We can extend this argument. The mass of the heart is proportional to the body mass, $M_h \propto M_b$, and we assume that the specific oxygen consumption of heart muscle is proportional to the specific oxygen consumption of the whole body: $\dot{V}_{O_2h}^* \propto \dot{V}_{O_2b}^*$. The conclusion is that, irrespective of the size of the animal, it should cost all mammalian hearts the same amount of oxygen to supply the body with 1 ml of oxygen. Another way to express this conclusion is that a constant fraction of all oxygen used is used by the heart for pumping the oxygen.

Will this hold during exercise as well as during rest? The available information is inadequate, and we can only suggest a possible answer. In humans, the blood pressure during exercise may increase by about one-half, to 180 or 200 mm Hg. The cardiac output, however, is not increased in proportion to the increase in oxygen consumption. This is because the amount of oxygen the muscles remove from the blood during heavy exercise is greatly increased. The oxygen content of mixed venous blood is therefore decreased, and the overall arteriovenous oxygen difference may be increased perhaps threefold. The result is that a rate of oxygen consumption during exercise at 15 times the resting level may lead to no more than about a fivefold increase in cardiac output. If the work of the heart were simply the product of pressure and volume, the work of the heart relative to oxygen transport might in fact be decreased during heavy exercise. This may not be true, however, because with the increased blood flow, the critical Reynolds number[1] for the transition to turbulent flow in the aorta may be exceeded, thus increasing the work of pumping. The importance of turbulent flow in exercising humans has not been well established, and not at all in other mammals. It certainly needs more attention, especially from the viewpoint of scaling.

Vascular turbulence

The question of vascular turbulence is quite interesting, but again, the available information is inadequate. It seems that turbulence in the bloodstream would be very wasteful. Nevertheless, it is often said that in the normal vascular system in humans at rest, turbulence is close to developing; thus, during exercise, turbulence must be of considerable importance.

The critical Reynolds number (Re) for the development of turbulence in a fluid flowing in a straight tube is said to be 2000; if Re exceeds this value, turbulence will develop.

The Reynolds number that applies to the flow of a homogeneous, Newtonian fluid in a straight rigid tube is calculated as follows:

$$\text{Re} = \frac{u\,d\,\rho}{\eta}$$

in which u is the mean velocity of the fluid in centimeters per second, d is the diameter of the tube in centimeters, ρ is the density of the fluid in grams per cubic centimeter, and η is the dynamic viscosity in poise. For the cardiac output in humans, assume 5 liters/min, and for the cross-sectional area of the aorta, 4 cm^2 ($d = 2.2$ cm); we then find the mean velocity of the blood to be 21 cm/sec. The viscosity of blood at 37°C is about 0.03 poise, the density is close to 1.0, and the Reynolds number will therefore be Re $= (21 \times 2.2 \times 1)/0.03 = 1540$ (Folkow and Neil, 1971).

---

1 The Reynolds number is a dimensionless number that expresses the ratio between inertial and viscous forces in a moving fluid. Its usefulness in this context is that the flow in a fluid becomes turbulent at high Reynolds numbers.

This calculation indicates that the flow in the aorta approaches the critical Reynolds number for turbulence.[2] Because cardiac output during heavy exercise may be increased some fivefold, the velocity of the blood and thus the Reynolds number should be five times higher, or well above the critical Re.

This conclusion does not take into account that the equation applies to steady flow in rigid tubes. The pulsatile flow of the blood may protect against turbulence, for turbulence is not established instantaneously, it takes some time to develop. Furthermore, the flow of a non-Newtonian fluid such as blood in an elastic tube is very different from that of a Newtonian fluid in a rigid tube. Finally, the critical Reynolds number is by no means as precise as indicated by the number 2000; Reynolds himself found that instability generally occurred at Re from 10 000 to 12 000. More recent results indicate that the observed critical Re depends primarily on the extent to which disturbances can be eliminated, and critical values of Re of more than 40 000 have been reached (Rouse, 1961).[3]

Can we say anything about the possibilities for the development of turbulence in relation to body size? Within the wide limits of uncertainty, we can try the following argument. The density (ρ) and viscosity (η) of mammalian blood are, for these purposes, scale-independent, and we can write Re $\propto u d$.

The cross-sectional area of the aorta (A, in cm^2) was determined by Clark (1927) for mammals ranging from a 16-g mouse to a 75-ton Greenland whale, a 5-million-fold difference in size (35 species of mammals measured). The data yield the equation $A = 0.094 M_b^{0.82}$ (with 95% confidence limits on the exponent of ±0.03). If we assume that cardiac output is proportional to metabolic rate, we obtain the mean velocity of the blood (u) as

$$u \propto \frac{\dot{Q}}{A} \propto \frac{M_b^{0.75}}{M_b^{0.82}} = M_b^{-0.07}$$

This suggests that the mean velocity of the blood decreases slightly with increasing body size. From the expression for the cross-sectional area of the aorta, we obtain that the diameter (d) is proportional to $M_b^{0.41}$. We can therefore calculate the change in Re in relation to body size:

2 In other parts of the vascular system, both blood velocity, u, and tube diameter, d, are lower, and therefore Re will also be lower.

3 There is a definite lower value of Re below which turbulence will not occur. In circular tubes this limit is about 2000, and below this critical limit the influence of viscosity predominates and stabilizes the flow so that turbulence does not occur. However, the upper critical limit for Re is indeterminate, and we should also note that turbulence develops at some distance from the entry to the tube and that it takes some time to develop.

$$\text{Re} \propto u \cdot d \propto M_b^{-0.07} \cdot M_b^{0.41} = M_b^{0.34}$$

This approximate calculation suggests that the value of the Reynolds number for blood flow in the aorta increases roughly with the third root of the body size. Evidently, if turbulence is not critical in the circulation in humans, it will be less so in any smaller animal. To what extent it will be important in animals much larger than humans, such as in horses, especially during heavy exercise, remains to be studied.

The instantaneous velocity during injection of blood into the aorta is higher than the mean velocity, but on the other hand, the duration of the pulse becomes shorter with higher heartbeat frequency during exercise. Furthermore, turbulence does not develop instantaneously; it takes some finite time before turbulence begins.

What remains of these considerations is that the Reynolds number increases with body size, but the diameter of the aorta is scaled out of proportion to the linear dimensions of the animal. This decreases the linear velocity of the blood and contributes to a reduction in the Reynolds number for the larger animals.

Circulation time

The blood volume for an average man is about 5 liters. At rest, the cardiac output (cardiac minute volume) is about 5 liters per minute. Therefore, the average time it takes for the entire volume of blood to make one full circuit and return to the heart is 1 min. Of course, all blood does not take exactly this time; some routes will be faster and some slower. What we consider is the average time for the blood to leave the heart, pass through the pulmonary circulation, return to the heart, leave again, pass through the systemic circulation, and finally return to the heart.

The way the cardiac minute volume is determined is to measure the rate of oxygen consumption (for a 70-kg man at rest, about 250 ml O_2 per minute) and the arteriovenous difference in oxygen concentration (about 50 ml O_2 per liter blood). To supply 250 ml of oxygen in a minute, we must therefore circulate 5 liters of blood; that is, we have obtained the cardiac minute volume. In general, cardiac output at rest is about 20 times the rate of oxygen consumption.

During exercise, the rate of oxygen consumption may be 15 times as high, but we do not need to increase the cardiac output 15-fold, because during exercise the arteriovenous oxygen difference may increase 3-fold. We can therefore manage with a 5-fold increase in cardiac output, and the mean circulation time during heavy exercise for a well-trained athlete may be 12 sec.

Table 11.4. Estimated blood circulation times for some mammals, calculated from the equation given in the text.

Mammal	Body mass (kg)	Circulation time (sec)
Elephant	4000	140
Horse	700	90
Human	70	50
Rat	0.2	12
Mouse	0.03	7
Shrew	0.003	4

How can we scale circulation time relative to body size? We obtain circulation time (t, in sec) by dividing blood volume (V_{blood}, in liters) by cardiac output ($\dot{Q}_h$, in liters/sec). We have previously used the equation for blood volume scaled to body mass (M_b, in kg), and cardiac output can be calculated from the rate of oxygen consumption by multiplying by 20, as described earlier. The factor 1/3600 is introduced to transform the hourly rate of oxygen consumption to the rate per second, and we obtain the equation

$$t = \frac{V_{blood}}{\dot{Q}_h} = \frac{0.0655 \, M_b^{1.0}}{20 \cdot 0.676 \cdot (1/3600) \cdot M_b^{0.75}} = 17.4 \, M_b^{0.25}$$

This gives circulation time as a function of body mass for mammals. The mass exponent is 0.25, which is to say that circulation time increases with increasing body size to the same extent that specific metabolic rate decreases with increasing body size (mass exponent $= -0.25$). The proportionality coefficient, 17.4, says that for a mammal of 1 kg body size, the average circulation time at rest is 17.4 sec. For a 70-kg human, the circulation time calculated from this equation will be 50 sec, a quite realistic number.

Let us calculate the circulation times for a range of body sizes, as in Table 11.4. We can see that an elephant will have a circulation time of well over 2 min, more than twice that for humans. The small animals show more dramatic differences. A 3-g shrew should have a circulation time of no more than 4 sec. In fact, the circulation time in shrews (and in hummingbirds, for that matter) may be even lower, because, as we saw earlier, their metabolic rates tend to be higher than would be estimated on the basis of body size alone.

During exercise, cardiac output increases and circulation time decreases. We said that the circulation time in a well-trained human athlete might be one-fifth of that at rest (the exact value varying with the arteriovenous oxygen difference). Accordingly, in the smallest shrew, the circulation time during maximal activity may be as low as 1 sec. In this amazingly short time the blood must leave the heart, pass through the lungs, go back to the heart, be pumped out through the body, and back to the heart again. No wonder shrews look a bit nervous most of the time! In the next chapter we shall return to the meaning of time for small and large animals.

Non-scaleable variables

We have already used information about several parameters that are important for the peripheral circulation. Some of these can be considered as non-scaleable physiological variables, in the sense that they are relatively constant and do not vary systematically with body size. Of course, they are not true "constants" in the sense of physical constants, such as atomic weights or the gravitational constant.

Examples of such non-scaleable physiological variables include the following: the viscosity of blood, plasma protein concentrations, hematocrit (volume of red cells as percentage of blood volume), blood pressure, mammalian red cell size, capillary diameter, and so forth. Perhaps the most surprising thing here is the relative constancy of red cell size and capillary diameter, two parameters that obviously are closely linked (although it is difficult to suggest which of the two may be the primary consideration). The problem is related to the constancy of cell size in general, an equally puzzling situation.

The relative constancy of red cell size was discussed in Chapter 10. Capillary diameter is highly variable, from zero to a diameter that will accommodate the passage of the red cells; thus, maximum functional capillary diameter can be said to be a non-scaleable variable. Because the oxygen consumption of a unit volume of tissue scales inversely with body size, we find some compensation in a higher capillary density in the highly active tissues of the small animal (i.e., a shorter diffusion distance between capillary and the oxygen-consuming cells). As was discussed earlier, the diffusion distance in muscle, the only tissue in which it has been adequately studied from the viewpoint of scaling, does not change with body size as regularly as might be expected.

Blood pressure, which also appears to be a scale-independent physiological variable, should not be thought of as merely being related to the circulation of blood. The role of the blood as a hydraulic fluid is of major importance, for it provides the force for ultrafiltration, important for the exchange of fluid and solutes across the capillary wall, and especially for ultrafiltration in the glomerulus of the kidney. In this context, the molecular size of the plasma proteins and their concentration also seem non-scaleable; thus, the capillary wall should, as an ultrafiltration device, have the same properties irrespective of body size. Perhaps the constancy of dimensions at the capillary level is directly related to these circumstances.

The viscosity of the blood is another scale-independent physiological variable. It could not be expected that viscosity should vary systematically with body size. The relative viscosity of blood (viscosity relative to water as unity) measured in tubes of millimeter magnitude is approximately 4 to 5. Some of this viscosity is due to the plasma proteins (plasma alone having a relative viscosity of about 1.7), but the major part is due to the red blood cells. As was mentioned earlier, the normal hematocrit for mammals is around 35 to 45 and is a non-scaleable variable, apparently adjusted to the maximum transport capacity for oxygen while minimizing the work required for pumping. With a constant plasma protein concentration and an optimized hemoglobin concentration, the viscosity cannot be expected to vary systematically with body size.

In summary, it is easy to understand some of the non-scaleable variables, such as blood pressure, blood viscosity, and hemoglobin concentration; what is more puzzling and still lacks an adequate and rational explanation is why red cell size and capillary diameter, within each major group of vertebrates, are body-size-independent.

12

The meaning of time

Small animals rush through life at a much faster pace than large animals: They breathe faster, their hearts beat more quickly, they move their legs faster – everything is speeded up. Can our clock time have the same physiological meaning for a large animal and a small animal?

The heart of a shrew beats 1000 times each minute, and the elephant's heart perhaps only 30 times. It takes the elephant heart about half an hour to beat 1000 times, the number of beats the shrew's heart races through in only 1 min. The same holds for other physiological functions: The shrew lives a much faster life than the elephant, and a time unit on the clock means something very different for these two animals. Obviously, physiological time is a relative concept, and the size of an animal dictates its time scale.

Time and frequency: How fast beats the heart?

Because the small heart beats with a higher frequency, the duration of each heartbeat is shorter. Frequency and time are inversely related; that is, frequencies are reciprocals of time. Conversely, time is the reciprocal of frequency:

$$\text{frequency} = \frac{1}{\text{time}}$$

The empirical equation for heartbeat frequency (f_h) relative to body mass (M_b, in kg), traditionally given in beats per minute, is, according to Stahl (1967),

$$f_h = 241 M_b^{-0.25}$$

The time required for each heartbeat (t_h, in min), or the duration of one heartbeat, will therefore be

$$t_h = \frac{1}{241} M_b^{0.25}$$

Recalculating t_h to seconds as the time unit gives a proportionality coefficient of 0.249, and the equation is

$$t_h = 0.249 \, M_b^{0.25}$$

For unity body mass, $M_b = 1$ kg, the duration of the heartbeat will be 0.249 sec, or almost exactly one-quarter of a second. This is four heartbeats per second, or 240 beats per minute.

A physiological frequency for which a great deal of material is available is the respiratory frequency for mammals. According to Stahl (1967), the respiratory frequency for mammals is

$$f_{resp} = 53.5 \, M_b^{-0.26}$$

The reciprocal of this equation gives the duration of each respiration (t_r, in min) according to the equation

$$t_r = \frac{1}{53.5} M_b^{0.26} = 0.0187 \, M_b^{0.26}$$

We note that the body-mass exponents in the equations for heartbeat frequency and respiratory frequency are not significantly different. If we calculate the ratio between heartbeat frequency and respiratory frequency, we get

$$\frac{f_h}{f_{resp}} = \frac{241 M_b^{-0.25}}{53.5 M_b^{-0.26}} = 4.5 \, M_b^{0.01}$$

The residual body-mass exponent, 0.01, is insignificant, and we therefore have the generalization that, on the average, the heart beats 4.5 times for each respiration. This ratio is independent of body size and should, on the average, hold true for all mammals. We must recall, as usual, that these empirical relationships are averages and that any given animal may deviate from the generalization. Nevertheless, as a general rule, we can expect that for all mammals the heartbeat frequency at rest is some four to five times higher than the respiratory frequency.

Birds breathe more slowly than mammals (their tidal volume is larger), and the heart rate is also lower. The equations for these two functions for birds are (Lasiewski and Calder, 1971)

$$f_h = 155.8 \, M_b^{-0.23}$$

$$f_{resp} = 17.2 \, M_b^{-0.31}$$

The ratio between the two gives

$$\frac{f_h}{f_{resp}} = 9.0\,M_b^{0.08}$$

This equation tells us that birds have about twice as many heartbeats per respiration as mammals. For a 1-kg bird, the ratio will be 9.0, but because the residual body-mass exponent is rather high, we cannot disregard the possibility of a significant body size dependence. The exponent, 0.08, suggests that the frequency ratio should increase somewhat in larger birds, but whether or not this is significant depends on the confidence limits for the exponents and the population sampled.

The 95% confidence limits for the exponents were ±0.04 for respiratory frequency and ±0.06 for heart frequency (Calder, 1968). This means that a residual mass exponent of 0.08 cannot be considered significant. Until further data are available, we can only say that in birds in general, the heart beats some nine times, more or less, for each respiratory cycle.

Metabolic rate and metabolic time

The most important measure of how fast an animal lives is its metabolic rate. As we saw earlier, the specific metabolic rate or specific power (P^*, power per unit body mass, M_b) decreases with increasing size according to the equation

$$P^* \propto M_b^{-0.25}$$

Because time is the reciprocal of rate, metabolic time (t_{met}), or physiological time, changes with body mass according to

$$t_{met} \propto M_b^{0.25}$$

This is the same relationship that we saw for the heart rate. For a very small animal, each heartbeat takes only a tiny fraction of a second, and in a large animal the heartbeat takes much longer in real clock time. The same relationship holds for all metabolic rate processes: Physiological time, relative to clock time, increases with increasing body size. This relationship between metabolic time and real time has been carefully considered in an interesting review by Lindstedt and Calder (1981), who discussed the problems of biological time from a variety of viewpoints, including the consequences for many problems of animal ecology.

The concept of physiological time can be applied to all sorts of rate processes. Let us examine a few examples.

The glucose turnover rate ($\dot{G}$) in the mammalian organism is related to body size with the same exponent as metabolic rate (Ballard et al., 1969). The specific turnover rate for glucose ($\dot{G}^*$, in mg/min per kilogram body mass) varies with body mass (M_b, in kg) as follows:

$$\dot{G}^* = 5.59\,M_b^{-0.25}$$

The turnover time for glucose, the reciprocal of the turnover rate, will therefore be

$$t_G = \frac{1}{5.59}\,M_b^{0.25} = 0.179\,M_b^{0.25}$$

This equation says that the turnover time (t_G) for 1 mg of glucose in a 1-kg animal is 0.179 min or a little over 10 sec and that the turnover time increases with increasing body size.

A number of other physiological rates, such as the renal clearance of inulin (Edwards, 1975) and the half-life of drugs (Dedrick et al., 1970), are related to body size with the same or very similar exponents. In other words, the rates, and therefore turnover times, excretion times, plasma half-lives, and so on, are in many cases directly related to metabolic time and to physiological time in general. It is therefore reasonable to use the metabolic rate as a measure of physiological time and remember that real clock time means very different things in the lives of small and large animals.

Life: How long, how fast?

The life span for small animals is scaled to their fast pace; they do not live very long. However, as we shall see, small and large animals enjoy approximately the same span of physiological life.

Let us return to one of the functions we discussed earlier: the breathing rate. A 30-g mouse that breathes at a rate of 150 times per minute will breathe about 200 million times during its 3-year life; a 5-ton elephant that breathes at a rate of 6 times per minute will take approximately the same number of breaths during its 40-year life span. The heart of the mouse, ticking away at 600 beats per minute, will give the mouse some 800 million heartbeats during its lifetime. The elephant, with its heart beating 30 times per minute, is awarded nearly the same total number of heartbeats during its life. (An alert reader may already have calculated that he may be dead, because his heart, beating at a normal rate of 60 or 70 beats per minute, should have used up its allotted number in some 20 or 25 years. Luckily we live several times as long as our body size suggests we should.)

When it comes to animal life span, the information we have is quite unsatisfactory and in many cases unreliable. First of all, what is meant by life span? Should we use the mean life span for animals in nature, where predation and other hardships tend to cut life short? Should we use the maximum life span in nature, as we obtain it from bird banding and similar methods? Should we accept the normal life span in captivity, where adequate food is available and predation and disease are absent? Or should we use the maximum life span under ideal conditions in captivity? These questions have been discussed by several authors, for example, Sacher (1959), Mallouk (1975), and Lindstedt and Calder (1976, 1981).

The life span (t_{life}, in years) for mammals in captivity has been found to vary with body size (M_b, in kg) according to the equation (Sacher, 1959)

$$t_{life} = 11.8\, M_b^{0.20}$$

A corresponding equation for captive birds is (Lindstedt and Calder, 1976)

$$t_{life} = 28.3\, M_b^{0.19}$$

These two equations reveal some striking facts. First of all, as we know, life span increases with body size. Furthermore, the exponents are virtually identical for birds and mammals. However, when it comes to number of years, birds of a given size live much longer than mammals of the same body size. The ratio between the two coefficients, 28.3 and 11.8, is roughly 2.5; that is, birds tend to live more than twice as long as mammals of the same size.

The next question is whether the exponents, 0.20 and 0.19, are significantly different from the exponent in the equation for metabolic time, 0.25. This question is virtually impossible to answer at the present time. Confidence limits on the exponents for lifetimes were not given in the original studies, but even if that had been the case, the selection (sampling) of animals for observations regarding lifetime was not the same as the selection of animals used for the metabolic equation. Furthermore, life spans are much more variable than metabolic rates, and an arbitrary selection of animals used for the life-span equation may produce much greater uncertainty than reflected in the confidence limits calculated from the data. Arithmetic confidence limits can look very attractive, and yet, because of the sampling of biological material, statistical confidence limits do not necessarily reflect the reliability of the information.

Long life and big brains

We saw earlier that humans live much longer than would be expected on the basis of body size. We also know that humans have inordinately large brains in relation to body size. This has led to several studies of life span as it relates to brain size. It was suggested by Sacher (1959) and also by Mallouk (1975) that this relationship could have important implications, as follows.

If maximum life span is related to brain size, the correlation is much closer than for body size. This led to an intriguing suggestion by Mallouk (1975). He speculated that the brain cells may produce a substance that is vital in organizing body repair. At maturity, the mammalian brain does not grow any more, and the "vital substance" would no longer be synthesized. The individual would then be left with a supply of this hypothetical substance that the body would draw on to manage its maintenance. The supply would diminish during life, and when exhausted, the individual would die of "natural causes."

Birds, relative to their body size, have much smaller brains than mammals, but live much longer. However, this could be explained if birds were to begin adult life with a much greater relative supply of the hypothetical vital substance in their smaller brains.

In any event, the relatively long life for humans, three or four times as long as expected on the basis of body size alone, is well correlated with our brain size, which is roughly four times as great as in mammals in general. However, Calder (1976) pointed out that it is a fallacy to assume that a good correlation indicates a causal connection. Calder, in a tongue-in-cheek comment on the hypothetical vital substance, pointed out that any scientist knows that correlation does not establish cause and effect; and he went on to show that life span can be correlated to the size of the spleen. Mammals, he said, have larger spleens than birds and have shorter life spans. He then facetiously postulated the existence of a "splenic senescence secretion (SSS)." Mammals, having more of this SSS, therefore live shorter lives than birds, which have less of it.

Although Calder's remarks were not intended seriously, they resulted in a somewhat angry reply from Mallouk (1976), who pointed out that longevity is related to relative brain size with much greater precision than any of the other allometric relationships or any other relative organ weight. This is where the matter rests; hypothetical life substances remain exactly that: hypothetical.

Real time

Animals live in the real world and cannot escape from real time. The day-and-night cycle and the seasons of the year are the same for all of us, but these cycles have very different meanings for small and large animals. Consider the high specific metabolic rate for the small animal and its consequence: a great need for food. The important question concerns how much energy is available and the rate at which it is used. The ratio between these two, the energy divided by the rate of use, will give the endurance time:

$$\text{endurance time} = \frac{\text{available energy}}{\text{rate of use}}$$

Now consider how long the body reserves may last. There is usually some food in the stomach and gut between meals, but the primary energy store is the fat of the body. There is no reason to believe that the relative amounts of fat that can be stored are very different in small and large animals, and let us therefore assume that fat storage is proportional to body mass ($M_b^{1.0}$). The metabolic rate for the animal is proportional to $M_b^{0.75}$, and therefore the endurance time is proportional to the ratio between these, or

$$\text{endurance time} \propto \frac{M_b^{1.0}}{M_b^{0.75}} = M_b^{0.25}$$

This relation states that endurance time increases with the size of the animal, and for a small animal the endurance time will be very limited. The small animal must eat almost continually: A nocturnal mouse must eat enough food to get through the day, and a hummingbird must find enough nectar to carry it through the night. There is only one way around this dilemma: to decrease the metabolic rate. This is indeed what many of the smallest warm-blooded vertebrates do: During the time of the day when they are unable to feed, they go torpid. Their body temperatures drop, and as a consequence metabolic rate decreases and endurance time increases.

It is more difficult for small animals to span an entire unfavorable season, such as winter. The options are few, but well known. They can migrate to better climates, as many birds do. Small mammals cannot migrate over long distances, but they can stockpile food. However, a better solution is to stockpile body fat and combine this with a decrease in metabolic rate by going torpid, that is, to hibernate. In this way the endurance time can be made to last through the winter.

Large animals can more easily survive the winter; their endurance time is longer. Bears can sleep through the winter without any drastic decrease in body temperature or metabolic rate. Large animals, with their long endurance time, can also more easily surmount geographical barriers. The large whales deposit tremendous amounts of fat while at their seasonal feeding grounds, and they carry out extensive oceanic migrations with minimal or no feeding. What 1 day of clock time is for a 10-g mouse may correspond to 2 months of time for a 100-ton blue whale.

A cold look at time

As we have seen, the warm-blooded animals, mammals and birds, can stretch their reserves over longer periods of time by decreasing their metabolic rates, by going torpid. What about cold-blooded animals, frogs, fish, and all sorts of invertebrates? For these, time must have less meaning; it must be less of a constant concept because their metabolic rates are not constant or nearly so. Metabolic rate, and thus metabolic time, vary greatly with all sorts of external effects: feeding, locomotion, and, more than anything else, temperature. Cold-blooded animals simply do not have a more or less fixed resting maintenance metabolic rate the way mammals and birds do.

The consequence is that we cannot apply to cold-blooded vertebrates and to invertebrates the principles we have discussed here, except in a most general way. When their metabolic rates are high, time is short; when their metabolic rates are low, time is extended. At low temperature they may even become completely inactive or "hibernate," and thus bridge whole seasons of unfavorable conditions. In fact, in the resting stage, many organisms can stretch their reserves over extended time periods, which for some may even last for years.

This uncertainty of what metabolic rate (energy per unit time) stands for makes it difficult to define the physiological meaning of time for these animals. This, of course, also makes it impossible to sort out any scaling principles in the relatively exact way we did for mammals. At least, with our present understanding, scaling principles are not directly applicable when it comes to time and the size of invertebrates.

13

Animal activity and metabolic scope

Real animals were not meant to sit still and patiently let a physiologist measure their "basal" metabolic rates. Animals eat, drink, sleep, run, chase, mate, and play. When their physical exertion is maximal, parts of the system, such as lungs and heart, must perform at a maximal level. Therefore, the limits on maximal performance are much more informative about animal design and much more interesting than the resting or idling level. Think of a racing car or an airplane sitting still with the engine running; the idling speed gives little information about maximal performance.

Maximal performance

During heavy physical work, such as running at top speed, oxygen is taken up in the lungs at a maximal rate and diffuses into the red blood cells of the lung capillaries, where it binds to the hemoglobin. The heart pumps the blood to the muscles, where oxygen diffuses from the capillaries to the cells and the mitochondria, which serve as the final oxygen sink. At each point in this chain, the flow rate of oxygen must equal the rate at which it is consumed in the sink.

Carbon dioxide, produced at a rate corresponding to the oxygen consumed, traverses the same pathway but in the opposite direction. At each step, the flow rate must equal the rate of production as CO_2 flows from the mitochondria into the capillaries, is circulated to the lungs, diffuses into the alveoli, and is dumped to the outside atmosphere.

Obviously, the capacity of each step in the gas transport system must be great enough to meet the maximal need; if a single link in the chain has a lower capacity, it will determine the rate for the entire system. On the other hand, if a single link in the system had a higher capacity than the others, it would never be called on to perform at that level, and there is no reason to believe that the organism would maintain an excess that could never be put to use.

An important principle

When we consider the scaling of the oxygen supply during maximal performance, we must be familiar with the principle of symmorphosis formulated by Taylor and Weibel (1981). This principle states that each functional system is adjusted to the level required for maximal performance and that no more structure is formed and maintained than is required to satisfy the maximal need (see Chapter 9).

The general concept of symmorphosis should apply to all levels of biological organization, because building and maintaining a capacity that will never be needed is expensive and wasteful. Therefore, no more structure than required should be formed and maintained.

This principle was developed in the introduction to a series of studies in which the morphological parameters of the respiratory system and the maximal rates of oxygen consumption were compared in wild and domesticated mammals ranging in size from 0.0072 to 263 kg. We shall soon return to these important studies.

Metabolic scope

The increase in metabolic rate between rest and maximal exertion is called the scope for metabolic activity, or *metabolic scope*. We are here concerned with the maximal rate at which oxygen is used and the maximal capacity of the oxygen supply system, the aerobic scope. The simplest way to express how the maximal rate of oxygen consumption relates to the resting rate is to consider how many times higher the maximal rate of oxygen consumption can be. This ratio, between $\dot{V}_{O_2max}$ and $\dot{V}_{O_2rest}$, is known as the factorial scope for oxygen consumption, or the *factorial aerobic scope*.

The factorial scope for oxygen consumption has been discussed by many authors. Hemmingsen, in his monumental monograph on scaling of metabolic rates, suggested that the scope for activity of mammals is about 10 times the resting rate and that it is independent of body size (Hemmingsen, 1960). However, it is well known that humans (Saltin and Åstrand, 1967) and horses (Brody, 1945) may reach much higher ratios, as high as 20 times the resting rates. A similar high factorial scope applies to dogs (Young et al., 1959). Why do these animals have exceptionally high scopes? Do they deviate from a general pattern because they are highly selected? Probably yes. The data for humans refer to maximally trained world-class athletes, and horses and dogs have for thousands of years been bred selectively for maximal performance.

Many small mammals, on the other hand, have factorial aerobic scopes of less than 10. Several investigators have been able to induce small rodents to reach peak $\dot{V}_{O_2}$ values six or eight times their resting rates, but not any higher. This is true not only for laboratory mice, rats, hamsters, and guinea pigs (Pasquis et al., 1965, 1970) but also for small wild rodents such as *Peromyscus* (Segrem and Hart, 1967) and *Microtus* (Jansky, 1959). The question that immediately arises is whether true maximal performance was measured or whether the animals merely were unwilling to run faster. Or do small mammals in general have lower scopes than large animals? Or are these animals distinct in the sense that in their normal lives they are never called on to run for long periods of time at a steady maximal rate?

One answer is certain: Small mammals are not necessarily restricted to the relatively low scopes that have been reported for small rodents, for bats can sustain rates of oxygen consumption 2.5 to 3.0 times greater than the highest metabolic rates that exercising terrestrial mammals of similar size can reach (Thomas, 1975). True enough, bats are highly specialized flying animals, but nevertheless, their flight performance shows that the mammalian organism is fully capable of sustaining such high rates; in other words, the lower factorial scopes found in rodents do not indicate a design limitation on the mammalian organism as such.

However, our main question remains unanswered: Is the factorial scope for activity of mammals independent of body size? The best answer that we presently can give to this question is found in the extensive studies of Taylor and Weibel (1981).

Taylor and Weibel

When Taylor and Weibel (1981) developed their plans to study the scaling of the entire respiratory system, from the diffusion of oxygen in the lung to the oxygen sink of the mitochondria, they decided that it was necessary to compare the morphological parameters and the maximal rates of oxygen consumption in the same individual animals.

The procedure for determining maximal rates of oxygen consumption was the same as Margaria (1976) had developed for humans. The animal ran on a treadmill at a sustained constant speed, and the rate of oxygen consumption was measured. When the treadmill speed was increased, the $\dot{V}_{O_2}$ increased up to a maximal rate, and further increases in speed could

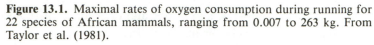

Figure 13.1. Maximal rates of oxygen consumption during running for 22 species of African mammals, ranging from 0.007 to 263 kg. From Taylor et al. (1981).

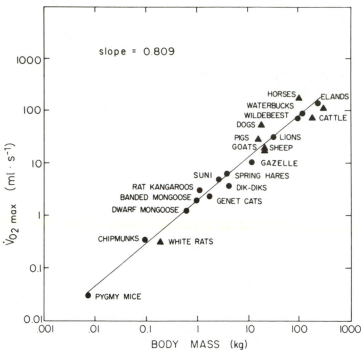

be achieved only by increased utilization of anaerobic glycolysis, as indicated by the accumulation of lactic acid in the blood. This would be the maximal $\dot{V}_{O_2}$ for that individual animal.

The $\dot{V}_{O_2 max}$ values for 14 species of wild mammals and 8 species of domesticated animals obtained with the treadmill method are plotted in Figure 13.1. The regression line for all the points had a slope of 0.809. For the wild species alone, the slope was 0.790; for the domestic species alone, it was 0.855. The coefficients and exponents in the regression equations for these relationships are given in Table 13.1. The confidence limits for the exponents (slopes) as well as the coefficients (intercept at unit body mass) reveal that there are no statistically significant differences between the three groups of animals. For the combined group of 22 wild and domesticated animals, the 95% confidence limits on the exponent were 0.747 and 0.870, a range that includes the accepted exponent for resting metabolic rate (0.75).

Table 13.1. Coefficients and body-mass exponents for maximal oxygen consumption in the equation $\dot{V}_{O_2 \max} = a M_b^b$, where M_b is body mass in kg (Taylor et al., 1981).

Animal group	Coefficient a	Mass exponent b
Wild, 14 species	1.94	0.790
Domesticated, 8 species	1.69	0.855
Wild and domesticated, 22 species	1.92	0.809

The purpose of these studies was to compare the maximal rate of oxygen consumption with morphometric determinations of diffusing capacity, both measured on the same individual animal. Taylor and Weibel decided to consider the 14 species of wild mammals, and for these the regression line was ($\dot{V}_{O_2}$ in ml O_2 sec^{-1}; M_b in kg)

$$\dot{V}_{O_2 \max} = 1.94 M_b^{0.79}$$

For comparison, Kleiber's equation (1961), recalculated to the same units, is

$$\dot{V}_{O_2 \text{std}} = 0.188 M_b^{0.75}$$

These equations, as well as the data in Table 13.1, suggest that there is a fairly constant factorial metabolic scope of about 10 in mammals in general and that this scope is independent of body size.

We should remember that this generalization in no way can be applied to any single animal species. For example, horses and dogs have factorial scopes that consistently are more than three times greater than those for cattle or sheep of the same body size. Thus, within a given size class, domestic animals may provide both high and low extremes of maximal metabolic capacity. This is the reason that including or excluding domestic animals has relatively little effect on the scaling exponent, but it widens the statistical confidence interval for both the exponent and the scaling coefficient.

The exceptionally high factorial scope, about 30 for domestic dogs, led to a further study of canids, both wild and domestic (Langman et al., 1981). The question was whether the high aerobic scope of the domestic dog is a result of intensive breeding for hunting and thus for top performance or whether it is a characteristic of other members of the dog family. The observed data for $\dot{V}_{O_2 \max}$ are listed in Table 13.2. As before,

Table 13.2. Maximal rate and factorial scope for specific oxygen consumption in four species of canids. All of these species have factorial scopes for oxygen consumption two to three times higher than for mammals in general (from Langman et al., 1981).

	Body mass (kg)	$\dot{V}_{O_2 \text{max}}^*$ (ml O_2 sec^{-1} kg^{-1})	Factorial scope for oxygen consumption
Gray fox	4.7	3.05	24
Coyote	12.4	3.07	31
Timber wolf	23.3	2.62	31
Dog	25.3	2.67	32

$\dot{V}_{O_2 \text{max}}$ was defined as the limit where no further increase in oxygen consumption could be obtained by increasing running speed, and any further increase in speed could be accounted for by lactate formation.

The results indicate that a high factorial metabolic scope is characteristic of wild as well as domestic members of the dog family. The approximately 30-fold increase in rate of oxygen consumption is in sharp contrast to the 10-fold factor found for most other mammals. Thus, the allometric signal for mammals in general is that their factorial metabolic scope is about 10, and the secondary signal is that canids deviate from this general mammalian pattern by a further factor of 3, giving a factorial scope of about 30.

Birds and bats

Virtually all we know about oxygen consumption in birds during flight has been obtained quite recently. Observations on birds in steady-state flight are very difficult to obtain, and even now data are available for only a few species. When data were reviewed by Berger and Hart (1974), only 11 species could be included. The least-squares regression equation for their data was ($\dot{V}_{O_2}$ in ml O_2 sec^{-1}; M_b in kg)

$$\dot{V}_{O_2 \text{flight}} = 2.43 \, M_b^{0.72}$$

This equation includes three hummingbirds measured during hovering flight, but Berger and Hart stated that the regression line would be similar by using only the species for which sustained forward flight had been observed.

This equation can be compared with the resting metabolism for birds, established for nonpasserine birds on a large amount of material (Lasiewski and Calder, 1971). In the same units as before, the equation is

$$\dot{V}_{O_2 rest} = 0.188 \, M_b^{0.72}$$

Because the exponents in the two equations are identical, the ratio between the coefficients gives the factorial scope 12.9. We should remember that the measurements used by Berger and Hart represented horizontal or hovering flight. However, the metabolic cost of horizontal flight does not represent the highest possible power output that birds are capable of, for birds can fly on an ascending path. This is intuitively obvious, and measurements on birds in sustained ascending flight have shown it to be true. Therefore, data from horizontal flight do not give the maximal factorial scope.

In any event, the equation for flight metabolism is based on too few observations, on different types of birds, on different flight patterns, and on different techniques. In fact, theoretically, the cost of bird flight, estimated for the speed at which the cost of transport should be at the minimum, should be more directly related to the body mass than is indicated by the exponent 0.72. The various theories behind this conclusion were extensively discussed by Tucker (1973), who also developed a practical simplification for estimating the power input during flight for birds of various body size. At the present time it is difficult to evaluate the discrepancies between theory and the meager data we have.

Observations on maximal metabolic rates for bats and birds were compiled by Thomas (1975). The data for three species of birds and two bats are given in Figure 13.2. This graph shows that bats, which are mammals, are capable of metabolic rates of the same magnitude as birds of the same body size. The least-squares equation fitted to the data combining bats and birds in Figure 13.2 is

$$P_{i, max} = 58.21 \, M_b^{0.65}$$

where power input, P_i, is in watts, and M_b is in kilograms. Recalculated to the units used earlier (ml O_2 sec^{-1}), the equation for flying bats and birds becomes

$$\dot{V}_{O_2 max} = 2.90 \, M_b^{0.65}$$

The confidence limits were not published by Thomas, but the location of the points in Figure 13.2 suggests that the confidence limits on both coefficient and mass exponent are narrow. Nevertheless, we must remember that statistically narrow confidence limits do not necessarily imply biologically equally narrow confidence limits. When the number of points is small, statistically significant data may be misleading in regard to biological validity.

Figure 13.2. Maximal metabolic rates measured in flying bats (crosses) and flying birds (circles), compared with running mammals. From Thomas (1975).

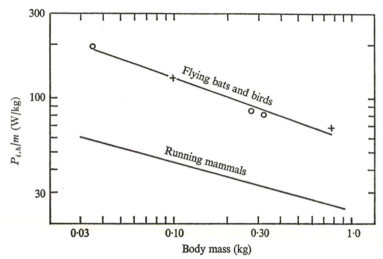

To summarize what we have learned, it seems reasonable to say that maximal metabolic rates for birds are more or less parallel to their resting or maintenance rates. In other words, a factorial scope for maximal activity of roughly 15 times the resting rate seems reasonable, as indicated by the few data compiled by Thomas.

This relationship leads to an interesting conclusion. If the factorial scope is unchanged with body size, the maximal oxygen consumption (and sustained power output) increases with body size with the slope 0.72. The theoretical power requirement for flight, however, increases in direct proportion to body mass, i.e., a slope of 1.0 (see Tucker, 1973). The inevitable conclusion is that the two lines must cross at some body size at which the maximal power output will no longer suffice for the power required for sustained flight. This crossing point is evidently at about 12 kg body mass, the size of the largest birds capable of sustained forward flapping flight, such as the Kori bustard and the largest swans.

To supply oxygen for flight: lungs and heart

We have seen that flying bats use oxygen at rates similar to those for flying birds. Does this require any special adaptation of their respiratory and circulatory systems, or are they similar to those of other mammals?

Table 13.3. Respiratory parameters at rest for a large fruit-eating bat (0.777 kg) compared with estimated data for a nonflying mammal and a nonpasserine bird of the same body mass (Thomas, 1981).

	Bat	Mammal	Bird
Respiratory frequency, min^{-1}	44.8	57.2	18.6
Tidal volume, cm^3	9.8	5.9	10.0
Ventilation, $cm^3 \, min^{-1}$	436.7	308.1	232.7
Oxygen consumption, $cm^3 \, min^{-1}$	14.2	9.5	9.4
Oxygen extraction	0.16	0.15	0.19

The structure of bird lungs is radically different from mammalian lungs in that bird lungs permit a unidirectional throughflow of air, as opposed to the in-and-out flow in mammalian lungs. Is the structure of the bird lung a prerequisite for flight? Evidently not, for bats have perfectly normally appearing mammalian lungs. We saw earlier that birds have much larger tidal volumes and lower respiratory frequencies than mammals. What is the situation for bats?

Respiratory parameters for a large fruit-eating bat, the flying fox, *Pteropus gouldii*, are given in Table 13.3 (Thomas, 1981). The results clearly put the bat in the general mammalian range, with the exception that its tidal volume (9.8 cm^3) is considerably larger than expected for a typical mammal of the same size (5.9 cm^3). In flight, the tidal volume of the bat increased to 40.6 cm^3, which is similar to the value predicted for a typical bird of the same body size (44.2 cm^3). Otherwise, the values for the bat are remarkably similar to those for other mammals. The resting rate of oxygen consumption was somewhat higher than predicted, and in horizontal flight it was increased by a factor of 10.7. Horizontal flight is unlikely to indicate maximum power, and the factorial metabolic scope will therefore be higher. Unfortunately, how much higher is impossible to say from the available data.

The size of the heart of bats, on the other hand, is more similar to that of birds than mammals (Table 13.4). In five species of bats between 5 and 150 g, the heart was consistently larger than predicted from the general mammalian equation, and the relative size increased with decreasing body size. For the smallest bat, the 5-g *Pipistrellus,* the heart was twice as large as predicted for a standard mammal of that body size, and quite similar to that predicted for birds.

The bats and shrews clearly show that the size of the heart in the smallest mammals is increased beyond the "normal" mammalian size in

Table 13.4. Heart sizes for five species of bats; the relative heart size increases with decreasing body size, and all are higher than the average value of 0.6% for mammals in general (Jürgens et al., 1981).

	Body mass (g)	Heart mass (% of body mass)
Rousettus aegyptiacus	146	0.84
Phyllostomus discolor	45.2	0.94
Molossus ater	38.2	0.97
Myotis myotis	20.6	0.98
Pipistrellus pipistrellus	4.85	1.26

order to provide oxygen at the high rates these animals require. Why is this so? As discussed earlier, it appears that a maximum heartbeat frequency of about 1200 to 1300 beats per minute is an absolute limit; therefore, the cardiac output can be sufficiently increased only by increasing the stroke volume, i.e., the size of the heart.

Factorial scope: cold-blooded vertebrates

When considering cold-blooded animals, we must first of all remember that their metabolic rates change with the temperature and that they therefore do not have one "standard" resting metabolic rate. Consider the curves for resting and maximal metabolic rates in Figure 13.3. The area between the two curves represents the metabolic scope, which not only changes with temperature but also reaches its maximum at a lower temperature than the highest resting rate. If we want to compare resting and activity rates, a meaningful comparison can be made only at the same temperature. Furthermore, it seems more informative to consider the factorial metabolic scope, that is, the ratio between the maximal activity rate and the resting rate at any given temperature.

For the speckled trout, the factorial scope at five different temperatures (5, 10, 15, 20, and 24°C) was 6.0, 5.5, 4.0, 2.6, and 1.7, respectively (based on the data used in Figure 13.3). In other words, the factorial scope decreases drastically with increasing temperature, which complicates any comparison of different body sizes.

Factorial metabolic scopes for a number of reptiles were compiled by Wood and associates (1978). For seven species of lizards, the factorial scope varied between 2.8 and 8.6; for four species of snakes, between 2.5

Figure 13.3. Rates of oxygen consumption for active speckled trout at different temperatures compared with their resting or standard rate. The ratio of active rate to resting rate (the factorial scope for activity) changes with temperature. From Graham (1949).

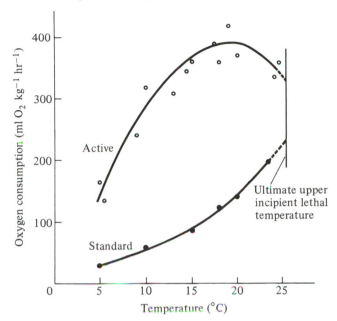

and 9.3; for four species of turtles, between 3.0 and 6.1. A fifth species (*Pseudemys scripta*) was listed with a very much higher factorial scope, 21.8. This last figure stands out as unique among reptiles, because the highest factorial scope otherwise was 9.3. The listed animals ranged in body size from 261 to 1637 g, with one exception, the giant tortoise, which weighed 100 kg (with a factorial scope of 6.1). There was no obvious relationship to body size in these data.

It is striking that factorial scopes of somewhere between 5 and 10 are not very unusual in reptiles. In fact, this range is quite similar to the factorial scopes that have been reported for many smaller mammals, especially rodents. However, because resting metabolic rates for reptiles are far below those for mammals of the same size, the metabolic scope expressed as the increase in the rate of oxygen consumption (given in ml O_2) will be a small fraction of that for mammals.

Let us return to the metabolic scope for fish. The salmon is an active fish and an excellent swimmer that has been studied extensively. The results are quite interesting, because they reveal a body size dependence.

First of all, temperature is more important for fish than for air-breathing animals. At higher temperatures, both the resting and activity rates are increased, and the oxygen available in the water may be a limiting factor. At high temperatures, the metabolic scope for activity may therefore be minimal. In swimming salmon, at temperatures above 15°C, fatigue sets in quite rapidly, because the cardiorespiratory system is incapable of delivering oxygen at the required rate (Brett, 1964). At 15°C the factorial scope for salmon is about 13, an impressive figure indeed, but it falls off rapidly with increasing temperature.

Is there a body-size relationship? An interesting result in Brett's studies (1965) was that the metabolic scope for salmon increased with increasing body size. For the smallest fish, 3.38 g, the factorial scope was 4.0, and it increased rather regularly to reach 16.3 at a body size of 1432 g. This change with increasing size reflects radical changes in the metabolic capacity as the fish grows. Shape and appearance differ little, but performance is drastically increased.

If we wish to summarize what we know about the metabolic scope for activity, we are in somewhat of a dilemma. For mammals, the factorial scope seems to be somewhat smaller for the smallest animals, but this is in part because several of the large animals (dog, human, horse) have exceptionally high factorial scopes. If we include bats, it is clear that small mammals are fully capable of very high factorial scopes, and we must conclude that for mammals, there is no clear body-size scale effect in the factorial scope. For reptiles, the available information is insufficient for any clear conclusion.

Finally, for the one species of fish that has been most carefully studied, the salmon, there is a clear scale effect in the factorial metabolic scope, which in the largest individuals is some four times higher than in the smallest specimens studied. Evidently, we need much more information carefully collected.

Muscle mass and muscle power

The power used in locomotion is provided by the muscles, and the muscles make up a substantial fraction of the body mass. Although the information we have about muscle mass is inadequate, it appears that the muscles make up roughly 40 to 45% of the total body mass of all mammals (perhaps except whales), irrespective of body size (Munro, 1969). This should suffice for an initial evaluation of scale; a more accurate estimate would be difficult, not only because few data are available but also because of the highly variable degree of fatness of different animals, particularly domestic animals used for meat production. A large amount of fat increases the total body mass, which distorts downward the percentage of muscle as compared with a more "normal" and less fat animal.

Information about birds is much more detailed, and the compilations made by Greenewalt (1962, 1975*b*) are particularly useful. The large pectoral muscles, the main flying muscles of birds, make up roughly 15% of the body mass. There are variations in this figure that are related to the flight habits of different birds. Good flyers tend to have larger pectoral muscles, but, on the whole, the percentage remains fairly constant over a wide range of body sizes.

It is interesting that the flight muscles of hummingbirds consistently make up a larger fraction of the body mass, some 25 to 30% as compared with 15% in other birds. This is in accord with the greater power requirement for hovering flight than for forward flapping flight (Weis–Fogh, 1972). Another interesting difference is that the muscles responsible for the upstroke of the wings make up about one-third of the total mass of the flight muscles in hummingbirds, whereas in other birds these muscles are only one-tenth of the total. Without any other knowledge, we might suggest that such a large muscle mass available for the upstroke indicates that lift is provided not only during downstroke but also during upstroke. This, of course, has been confirmed by aerodynamic analysis of wing movements of hovering hummingbirds (Stolpe and Zimmer, 1939). This is another example in which the allometric signal of scaling and the secondary signal provided by deviations from it provide important information about basic principles.

Let us turn to the power provided by the muscles. Power is work per unit time, and work, in turn, is force times distance. We should therefore first examine the force that muscles can exert and then the contraction distance. This gives us information about the work that the muscle can provide in a single contraction. Next comes the time parameter: the duration of the contraction event. How can we evaluate these three parameters and relate them to body size and scaling?

It appears that the maximum force, or stress, that can be exerted by any muscle is inherent in the structure of the muscle filaments. The maximum force is roughly 3 to 4 kgf/cm^2 cross section of muscle (300–400 kN/m^2). This force is body-size-independent and is the same for mouse and elephant muscle. The reason for this uniformity is that the dimensions of the thin and thick filaments, and also the numbers of cross-bridges between them, are the same. In fact, the structure of mouse muscle and elephant muscle is so similar that a microscopist would have difficulty identifying them except for a larger number of mitochondria in the muscles of smaller animals. This uniformity in maximum force holds not only for higher vertebrates but also for many other organisms, including at least some but not all invertebrates.

The shortening of vertebrate skeletal muscle is also body-size-independent. The maximum relative shortening, or strain, seems to be rather constant at around 0.3. From this we can conclude that the maximum work (force × distance) performed in one contraction, when calculated per unit volume of muscle, is also an invariable and is independent of size. This generalization was recognized by A. V. Hill (1950) and is well supported by later evidence.

The conclusion that the maximum work per contraction is scale-independent is fully in accord with our present knowledge of the structure of skeletal muscle. The number of filaments per cross-sectional area of muscle is the same (i.e., filament thickness is constant), and the sarcomere length, the length of the filaments, and thus the maximum overlap between thick and thin filaments, are all of the same magnitude in small and large vertebrates. This uniformity in filament thickness, length, and overlap explains the constancy of stress and strain (although it does not answer the question why these parameters remain scale-independent).

If the work per contraction is constant, the power output (work per unit time) during contraction will be a direct function of the speed of shortening, or the strain rate. For repeated contractions, as in running or flying, the average power output of a muscle will be directly proportional to the frequency of its contraction. We should note, however, that the extremely high frequencies of contraction of the wing muscles of some flying insects do not imply power output at an inordinately high level, for these muscles (because of the peculiarities of the flight apparatus) do not shorten by more than perhaps a couple of percent.

In summary, we can say that vertebrate skeletal muscle is uniformly of similar structure and that maximum force and shortening are scale-independent. In regard to speed of shortening, there are vast differences. If we compare analogous muscles, such as the muscles of locomotion, we find that the frequency of contraction increases regularly with decreasing body size. Hence, power output is scaled to increase with decreasing body size. In the next chapter we shall return to the details of how the power output of locomotory muscles scales with body size.

14

Moving on land: running and jumping

Running on land

We know that a mouse has shorter legs than a horse and must take more steps to cover the same distance on the ground. As a first approximation, consider a series of geometrically similar animals. The number of steps (n) the animal must take per unit distance moved will be inversely related to its linear dimension, or the one-third power of its mass. The relationship will be

$$n \propto \frac{1}{l} \propto M_b^{-1/3}$$

The assumption of geometric similarity is not as unrealistic as one might expect. The limbs of mammals from shrews to elephant were studied by Alexander and associates (1979a), who found that the mean length of the limb bones is related to the body mass with the exponent 0.35. The series covered 37 species of mammals and represented nearly the full size range of terrestrial mammals. In spite of the substantial differences in shape and body proportions between the largest and smallest mammals, the lengths of their limb bones came surprisingly close to geometric similarity. For complete geometric similarity to prevail, a 3-ton mammal would have linear dimensions 100 times the corresponding linear dimensions of a 3-g mammal; for an exponent of 0.35, the 3-ton animal would have linear dimensions 126 times those of the 3-g mammal.

Next, consider the frequency of leg movement, which depends on the speed at which the animal runs. At the moment, we have no a priori way of predicting the speed of an animal; we can only say that for two animals running at the same speed, the frequency of strides will be inversely proportional to the linear dimension, or to the one-third power of the body mass.

We can, however, make some reasonable predictions about the cost of running. As far as we know, all mammalian muscles can develop the same maximal force per cross-sectional area, about 3 to 4 kgf/cm^2. This maximum force should be independent of body size, because the force of contraction is provided by the cross-bridges between the thin and thick filaments. The number of cross-bridges is determined by the sarcomere length, which in mammalian muscle is constant, independent of body size. If corresponding muscles are isometric, then cross section is proportional to l^2, and length to l, and the muscle mass remains proportional to body mass.

There is a great deal of information indicating that the total muscle mass of mammals does not vary in any systematic way with body size. In many mammals the muscles make up some 40% of the total body mass, but it is difficult to evaluate whether substantial deviations from this mean are real or depend on how the muscle mass is determined, how fat the animal is, how carefully the dissection is carried out, and so on. The best material we have available is that from Alexander and associates (1981), who determined the size of the major leg muscles in mammals ranging from shrews to elephants. Although some animal groups tended to contain more muscle than others, and the distribution between proximal and more distal leg muscles differed, the total muscle mass tended to be proportional to $M_b^{1.0}$.

We know that most vertebrate muscles can shorten by the same fraction of their resting length. Because the work performed in a single contraction is the product of force and the distance over which the force is acting, it follows that the work performed in one contraction will be proportional to the mass of the muscle. With the muscle mass proportional to body mass, the work per step should therefore be proportional to body mass.

Returning to our model of isometric animals, we have that the work per step should be proportional to $M_b^{1.0}$. Because the number of steps per unit distance is inversely proportional to $M_b^{1/3}$, the work of running over one unit distance, say 1 km, should be

$$\text{work per km} \propto M_b^{1.0} \cdot M_b^{-1/3} = M_b^{0.67}$$

To compare small and large animals, consider the work of running per kilogram of body mass per unit distance, which we obtain by dividing the preceding expression by body mass:

$$\text{work per kg per km} \propto M_b^{0.67} \cdot M_b^{-1.0} = M_b^{-0.33}$$

This expression says that the work required to move 1 kg of animal over a given distance decreases with increasing body mass. This is a direct consequence of the many more steps required for the small animal to cover the same distance on the ground, each step requiring work in direct proportion to body mass.

This is the result of the simplest possible theoretical model. When we consider what real animals do as they run, we shall discover that our simplified arguments correspond amazingly well to the real world.

The energy cost of running

We have just learned that large animals, because they take fewer steps, should use less energy to move one unit of body mass over one unit of distance. Let us, for this purpose, consider the increment in energy expenditure caused by moving along the ground. The amounts of energy used by various mammals of different sizes when running were examined by Taylor and associates (1970). Some of the observations on mammals running at various speed are shown in Figure 14.1. The oxygen consumption, or energy expenditure, increased with the speed of running; this, of course, was expected. What was surprising was that energy expenditure increased linearly with running speed. We shall momentarily see that this is a great advantage in comparing the different animals.

What is clear from Figure 14.1 is that energy use (per unit body mass) increases much faster with running speed in the smallest animal (steeper slope of the regression line).

The slopes of the regression lines indicate the increment in energy use with an increment in running speed. In fact, the slope gives the cost of running, or, more precisely, the energy used to move one unit of body mass over one unit distance. Using the units on the coordinates, the dimensions of the slope are

$$\frac{\Delta y}{\Delta x} = \frac{\text{liter } O_2 \text{ kg}^{-1} \text{ hr}^{-1}}{\text{km hr}^{-1}} = \text{liter } O_2 \text{ kg}^{-1} \text{ km}^{-1}$$

This tells us that the slopes of the lines in Figure 14.1 actually represent the numbers of liters of O_2 used to move 1 kg of body mass over 1 km. In other words, the slopes give a weight-related mileage expression for the cost of running. It is analogous to considering the amount of fuel a vehicle uses to move 1 km relative to its mass; a train uses more fuel than an automobile, but relative to its mass it uses much less. It is more economical to be big, and this applies to animals as well.

Figure 14.1. Oxygen consumption for running mammals increases with the speed of running. For each species, the increase is linear, but the oxygen consumption increases more steeply for a small animal than for a large animal. From Taylor et al. (1970).

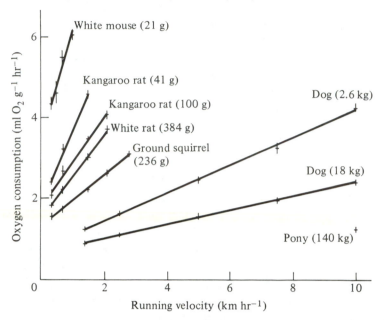

Because the cost of running, expressed in oxygen per kilogram per kilometer (the slope), for each animal is independent of the speed at which it runs, the speed has conveniently been eliminated as a variable in calculating the cost. We can therefore compare the cost of running for various animals, although they move at different speeds. We can let each run at a speed that is natural to it, and we can compare animals even if the ranges of speeds at which they run do not overlap.

How does the cost of running relate to the body size of the animal? Figure 14.2 gives the data obtained by Taylor and associates in their first publication (1970). This graph shows what we have already seen, that the cost of moving one unit of body mass over a given distance is lower for a large than for a small animal. The regression line in Figure 14.2 has a slope of −0.40, based on six small to medium-sized plus one large mammal (horse) obtained from the literature.

Taylor noted that human runners fell outside the regression line for four-footed mammals and wondered if the cost of bipedal running always differs from that for quadrupedal running. This question was

Figure 14.2. The cost of running, expressed as the oxygen needed to transport 1 kg of body weight over 1 km, decreases regularly with increasing body size. Data for man (bipedal running) fall above the line representing data for mammals running on all four legs. From Taylor et al. (1970).

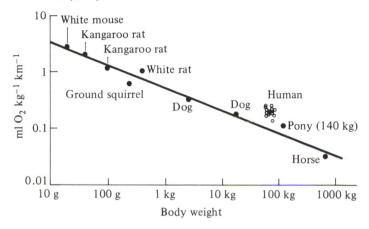

taken up by Fedak et al. (1974), who investigated seven species of birds. The regression line they reported was indeed different. It had a slope of −0.20, and the human data seemed to fit the same line.

This raised the question whether or not there is a fundamental difference between the scaling of bipedal and quadrupedal locomotion relative to body size. A few animals seem to move with ease either way, and Taylor and Rowntree (1973) trained chimpanzees (*Pan troglodytes*) and capuchin monkeys (*Cebus capucinus*) to run on a treadmill, either on two or on four legs. Both animals expended the same amount of energy, whether running on two or on four legs. However, this did not fully answer the question, because both were of intermediate size, and the earlier results indicated that there should be significant differences only at very small or very large body sizes.

The initial observations by Fedak et al. (1974) were based on a small number of species, and later reviews by Paladino and King (1979) and Fedak and Seeherman (1979) on a larger number of species showed no significant differences between birds and mammals, that is, between bipedal and quadrupedal running. This result gives us some satisfaction, because it seems to confirm that animals go about matters as economically as feasible, whether they run on two or four legs. It is now increasingly certain that the exponents calculated in the earliest compilations were not based on a sufficiently wide range of animals, and when a sufficient number of species is included, the exponent is close to −0.33 for both birds and mammals (Table 14.1).

Reptiles have resting metabolic rates 10-fold lower than mammals and birds of the same size. It may therefore seem surprising that the increment cost of running for lizards is similar to that for birds and mammals. For a series of lizards ranging from 14 to 1200 g, Bakker (1972) found that the cost of running (ml O_2 g^{-1} km^{-1}) decreased with body mass (M_b, in g) to the power 0.33. The equation for the straight regression line fitting his data is

$$\text{cost} = 5.9 \, M_b^{-0.33}$$

This equation is indistinguishable from that for later compilations that include larger numbers of mammals and birds, and it justifies the inclusion of reptiles in several of the studies listed in Table 14.1. The most interesting conclusions are (1) that the increment cost of running for lizards is the same as that for mammals of the same body size, although the lizards have a much lower and variable resting metabolic rate, and (2) that the cost of running for lizards decreases with increasing body mass to the one-third power, as it does for running mammals and birds.

If ants ranging in size from 3 to 36 mg are included in the calculation, the resulting equation is similar to those for mammals and birds (Jensen and Holm–Jensen, 1980). Is it a coincidence or an expression of a fundamental principle that ants fall on an extension of the vertebrate regression line? We do not know.

We have another important question to answer: Is it a fortuitous coincidence that the empirically determined exponent is −0.33, or does it serve to validate the simplified scaling considerations developed earlier? In any event, animals are not isometric; they have all sorts of different sizes and proportions. Some deviate substantially from the general pattern, but when we examine the broadest possible range of animals, it appears that similarities in animal locomotion nevertheless confirm an amazing consistency in energy cost.

If we look at this from another viewpoint, its evolutionary significance, the result is more reasonable. Presumably, economy with available fuel is an advantage, and we expect that animals live as economically as possible. Small and large animals appear to be up against identical constraints that cannot be exceeded. Muscle contraction is as economical as possible, and for each animal the economy of moving about is as economical as its body size permits. The small animal must take more steps to move a given distance, and this costs it more. If we relate the cost of

Table 14.1. Equations for the cost of running related to body size (M_b, in g) as obtained by various investigators; the mean of the body-mass exponents for all these equations is -0.32.

Study	Number of species	Equation for cost (ml O_2 g^{-1} km^{-1})	Cost for $M_b = 1$ kg (liters O_2 km^{-1})
Taylor et al. (1970)	7 mammals	cost $= 8.46\,M_b^{-0.40}$	0.53
Bakker (1972)	8 reptiles	cost $= 5.9\,M_b^{-0.33}$	0.60
Fedak et al. (1974)	7 birds	cost $= 2.45\,M_b^{-0.20}$	0.62
Cohen et al. (1978)	22 mammals	cost $= 6.35\,M_b^{-0.34}$	0.61
Paladino and King (1979)	52 mammals, birds, reptiles	cost $= 5.01\,M_b^{-0.32}$	0.55
Fedak and Seeherman (1979)	69 birds, mammals, reptiles	cost $= 3.89\,M_b^{-0.28}$	0.56
Jensen and Holm-Jensen (1980)	72 mammals, birds, reptiles, ants	cost $= 8.61\,M_b^{-0.352}$	0.76
Taylor et al. (1982)	62 mammals, birds	cost $= 4.73\,M_b^{-0.316}$	0.53

locomotion to one step (or to any other linear dimension), we find animals to be equally economical, whether large or small.

How fast animals run

We do not fully understand the relationship between animal size and how fast an animal can run. For one thing, animals are so very different; and some of those we know best, horses and dogs, have been bred for thousands of years for high speed and endurance. Furthermore, it is difficult to establish the maximum speed for any given animal, because when we are unable to make an animal run faster, is it because he cannot or will not run faster? Another open question concerns what is meant by maximum speed. Is it sprint speed, which is largely anaerobic, or maximum sustained aerobic speed? And if so, sustained for how long?

We saw earlier that there is a linear relationship between metabolic power (rate of oxygen consumption) and the speed at which an animal runs. As before, we shall consider only the increment in metabolic power caused by running, because at high speed the resting metabolic power is so small as to be nearly insignificant.

We can turn the linear relationship between speed and metabolic power around. Instead of saying that metabolic power increases linearly with running speed, we can say that running speed increases linearly with available metabolic power.

In the preceding section we saw that the cost of running per unit body weight decreases with increasing body size with an exponent of about 0.33. If instead we express the cost of running a given distance in terms of the entire animal, the cost increases with body size with the exponent 0.67 (both sides of the equation multiplied by $M_b^{1.0}$). This is indicated in Figure 14.3, which also shows lines with the slopes of 0.75 and 0.85.

We have seen that resting metabolic power for mammals increases with body size to the power 0.75. It has many times been suggested, although on weak grounds, that the maximum metabolic power would be approximately 10 times the resting. If this were so, the available maximum power (exponent = 0.75) would increase faster than the power requirement for running (exponent = 0.67). If the maximum running speed were a simple function of available power (which it is not), the running speed should increase with body size according to the difference between 0.75 and 0.67, that is, with the body size to the power 0.08.

When it comes to maximum power, or maximum rate of oxygen consumption for mammals, the available information is much better than for maximum speeds.

Figure 14.3. Comparison of slopes (without consideration of absolute magnitudes) for cost of running (slope = 0.67), for resting (slope = 0.75), and for maximal oxygen consumption (slope = 0.85).

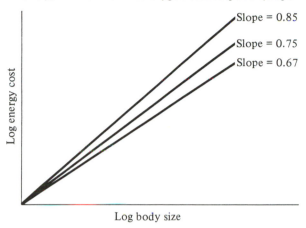

The maximum rates of oxygen consumption measured in 22 species of wild and domestic animals by Taylor and associates (1981) showed that maximum oxygen consumption increased with body mass to the power 0.81. When these authors included an additional 23 values taken from the literature, giving a total of 55 wild and domestic species, the slope of the regression line was 0.85. This suggests that maximum power increases with body size faster than the resting rate. Let us accept the highest figure, 0.85, for the available aerobic power. We said before that the cost of running should increase with the exponent 0.67. Available power therefore increases faster than the cost of running, the exponent for the difference being 0.18. Over the full size range for land mammals, from a few grams to a few tons, this would give a 10-fold increase in power relative to the cost of running, suggesting that there might be a 10-fold difference in running speeds as well.

We can now compare these somewhat theoretical arguments with observations made by Pennycuick (1975) on East African animals ranging over a 200-fold size range, from about 15 to 3000 kg. Pennycuick found that the stepping frequency for these animals decreased with body size; that is, large animals moved their legs more slowly than small animals. The most interesting aspect was that the slopes of the regression lines for mean stepping frequency were the same whether the animals moved in a walk, in a trot, or in a canter. For each gait, the stepping frequency was related to a linear dimension, the shoulder height of the animal, and the slope for each gait was within the 95% confidence limits of -0.50. If the stride length, also a linear measurement, likewise is proportional to body mass to the power 0.33, the stride frequency will be proportional to $M_b^{-0.17}$ ($M_b^{-0.50 \times 1/3}$). The speed (stride length × frequency) will therefore be proportional to body mass to the power 0.17:

$$\text{speed} = \text{stride length} \times \text{frequency} \propto M_b^{0.33} \times M_b^{-0.17} \propto M_b^{0.17}$$

These were animals moving in nature at their natural speeds, and it is quite clear that larger animals in the wild move faster than smaller animals. This does not exclude the possibility that maximum sprint speeds can be related to body size in a different way, but we lack adequate information. It is evident, however, that the argument of Hill (1950), that all animals should be moving at equal maximal speed, is most unlikely. Hill developed his argument theoretically and supported it with observations on racing speeds over a narrow size range, from whippets to horses. Both are extraordinary animals, bred for speed, and they have maximum rates of oxygen consumption between three and four times higher than those for goats and cows of the same body size (Taylor and Weibel, 1981).

Another attempt at comparing running speeds for different animals was made by Heglund and associates (1974). Because it is so difficult to establish the maximum speed of an animal, these authors suggested that the speed at the transition from trot to gallop can be considered as an equivalent speed for comparing animals of different sizes. This transition occurs at lower speed and at higher stride frequencies in smaller animals. Plotting the stride frequency at the transition point against the body size gave a straight regression line with the slope of -0.14 (animal size ranging from mouse through horse). The difference between this and Pennycuick's observations on East African wild animals (slope $= -0.17$) is insignificant.

We now have some idea of how running speed may change with body size. Considerations of the maximal power available for running, as well as empirical observations, suggest that running speed should increase with increasing body size with an exponent of about 0.17. A better analysis must await an improved understanding of how the energy is used in locomotion. The external work done by animals running horizontally is minimal, and most of the work is dissipated internally as heat. How this relates to body size provides a very complex set of problems.

Running uphill and carrying loads

Thus far, we have considered only animals running on the level. To run uphill requires additional work. A body that is moved uphill acquires potential energy, which is the product of its weight and the vertical distance it has moved. We can therefore expect that the work needed to move one unit of body weight over one unit of vertical distance should be the same for all animals, irrespective of body size.

The argument is as follows. To lift 1 kg of body weight 1 m vertically increases its potential energy by 1 kgf m (which equals 9.8 J or 2.34 cal, and corresponds to an oxygen consumption of 0.49 ml O_2). We assume that large and small animals have muscles that work with the same maximum efficiency, say 25%. The animals should therefore use four times the preceding amount, or 2 ml O_2 to move 1 kg body weight 1 m vertically, irrespective of body size. This expected similarity in the metabolic cost of moving vertically has been amply confirmed by many investigators for mammals ranging in size from mice to horses.

Some of the data on vertical locomotion were reviewed by Cohen and associates (1978). Several of the observations clustered around an increment cost of 1.36 ml O_2 per kilogram per vertical meter moved, without any clear correlation with body size. (The fact that this suggests an efficiency somewhat higher than expected should not be taken too seriously because the measurements refer to the increment in the cost, and the vertical component is not necessarily simply additive to the horizontal component.) The observation that the cost of moving vertically is independent of body size has an interesting consequence: It must be much easier for a small animal than for a large animal to run uphill. Why?

The answer is simple: At rest, a mouse has a specific metabolic rate about 15 times higher than a 1000-kg horse. Because the vertical component of moving one unit of body weight uphill is the same for the two animals, the increase in metabolic rate attributable to the vertical component, relative to the resting rate, will be only 1/15 as great in the mouse as in the horse.

This surprising conclusion was experimentally tested by Taylor and associates (1972). Mice were trained to run on a treadmill at a 15-degree incline, which is a rather steep hill. The rates of oxygen consumption for these mice were not significantly different whether they were running on the level or uphill or downhill. For chimpanzees (17 kg body weight), oxygen consumption nearly doubled on the uphill slope, as compared with running on the level. For horses, the increase caused by moving uphill, as compared with horizontally, is several times their resting metabolic rate (the precise amount depending on speed).

Taylor concluded that moving vertically at 2 km/hr requires an increase in oxygen consumption of 23% for a mouse (hardly noticeable), 189% for a chimpanzee (nearly doubling), and 630% for a horse (heavy work). This explains the ease with which a squirrel moves up and down a tree trunk, apparently without effort. For an animal of such small size, it makes little difference whether it runs up or down.

A related question is the cost of carrying an extra load. Taylor and associates (1980) measured the energy costs of carrying loads for rats, dogs, humans, and horses for loads ranging between 7 and 27% of the body mass. In these experiments the oxygen consumption of the animals increased in direct proportion to the added load. For example, if the load was 10% of the body mass, the oxygen consumption was increased by 10%, and so on, as shown in Figure 14.4.

The direct proportionality between increased oxygen consumption and added load has an important consequence. Because small animals have a higher cost of locomotion than large animals, small animals expend relatively more energy to carry a given load over a given distance. Carrying a load causes the same fractional increase in oxygen consumption for small and large animals, but in absolute terms the cost is relatively higher for the small animal.

Scaling of jumps

A discussion of animals moving uphill would not be complete without a consideration of jumping. We shall encounter two interesting facts: One pertains to the effect of body size on how high an animal can jump, and the other will reveal a necessary change in design, a discontinuity, as body size decreases.

A flea, which is less than 2 mm long, can jump to a height more than 100 times its body length. Even the best human athlete cannot jump much higher than his own body length. However, if we consider some simple rules of scaling, the seemingly poor performance of the athlete is by no means inferior to that of the flea.

To make the simplest possible argument, consider two animals of isometric build, the larger animal with all linear dimensions increased in exactly the same proportions. What is the effect of scaling on the expected height to which the larger animal can jump? The answer is that, irrespective of body size, large and small animals should be able to jump to equal heights.

Figure 14.4. Effect of carrying a load on the energy cost of running. The coordinates express the ratio between the measured variables for a loaded animal versus an unloaded animal. Each point represents the mean of measurements made at one single running speed. Vertical bars represent ±2 SE of the ratio of oxygen consumption while the animal carried a load to its oxygen consumption without a load at the same speed. The regression line (slope = 1.0) represents identity of the ratios on the two axes. From Taylor et al. (1980).

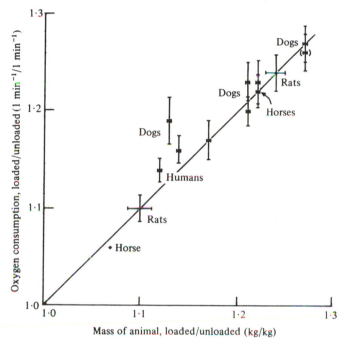

Assume that the jumping muscles make up the same fraction of the body mass in the two animals, a realistic assumption. The muscle force is proportional to its cross-sectional area, and the shortening is proportional to the initial length of the muscle. The cross section times the length is the volume of the muscle, and the energy output of a single contraction is the product of force and distance. Energy output is therefore proportional to the muscle mass and, in turn, to body mass.

In a jump, an animal uses a single contraction of the jumping muscles, and the energy available for the takeoff and used for acceleration is therefore the same relative to the body mass. The same amount of energy (work) per unit mass can lift the bodies to the same height and no more. The conclusion is that similar animals of different masses should jump to

Table 14.2. Jumping performances of four animals that differ in body size by a factor of more than 100 million.

	Flea (*Pulex*)	Click beetle (*Athous*)	Locust (*Schistocerca*)	Man (*Homo*)[a]
Body mass	0.49 mg	40 mg	3 g	70 kg
Height of jump	20 cm	30 cm	59 cm	60 cm
Acceleration distance	0.075 cm	0.077 cm	4 cm	40 cm
Takeoff speed	190 cm/sec	240 cm/sec	340 cm/sec	343 cm/sec
Takeoff time	0.00079 sec	0.00064 sec	0.00235 sec	0.233 sec
Acceleration[b]	245 g	382 g	15 g	1.5 g

[a] Estimate based on center of mass lowered to 60 cm above ground at beginning of jump, accelerated over 40 cm to 100 cm at takeoff, and lifted to 160 cm. World record for standing jump is about 165 cm above ground.
[b] Acceleration is expressed relative to the earth's gravitational acceleration, *g*.

the same height, provided that their muscles contract with the same force.

When it comes to actual animals, the situation is not far from the theoretical. The jumping heights recorded for four very different animals are listed in Table 14.2. We consider only standing jumps, for a running jump utilizes the kinetic energy of the running animal to increase the height of the jump. A man can clear about 1.6 m in a standing jump. His center of mass, however, is not lifted over this distance; it is lifted over less, because it is not at ground level when he jumps, but rather at about 1 m. Hence, the height of the standing jump for the man is only 60 cm. The body masses of the animals in Table 14.2 differ by a factor of more than 100 million, but, nevertheless, the heights of their jumps range over no more than a three-fold difference.

The jump of the flea would, in fact, be even higher if air resistance were not a consideration. Because of its small size (high surface-to-volume ratio), the air resistance is appreciable, and if the same amount of energy were imparted to a flea in vacuum, its jump would be about twice as high.

Air resistance is very important for the small animal. Streamlining a flea will have no effect on its air resistance because of the boundary layer of air that it drags along. However, air resistance changes with speed, and therefore also with the height of the jump (because the initial speed is higher for a higher jump). For this reason it is not possible to state simply the role of air resistance, but experiments in which insects were projected vertically by a spring gun have been very informative (Bennet–Clark and Alder, 1979). The ratio between the height reached in air relative to the height reached in vacuum (in the absence of air resistance) can be called the jump efficiency. If a flea theoretically could jump to 2 m in a vacuum, it would reach only 0.53 m in air, giving a jump efficiency of 0.27. In this case, the air resistance consumes three-quarters of the energy used for the jump (Table 14.3).

For a 440-mg grasshopper with a jump efficiency of 0.74, only a quarter of the energy going into the jump is used to overcome air resistance. The galago, a small primate that probably holds the world record for a standing jump because it is able to jump more than 2 m vertically, has a jump efficiency of 0.97. In other words, only 3% of the energy going into the jump is needed to overcome air resistance. For even larger animals the air resistance in a jump is completely insignificant.

How is it possible for the galago to make standing jumps more than three times as high as a human athlete? Under well-controlled conditions,

Table 14.3. Jumping performances calculated for three different size animals, based on an assumed jumping height in a vacuum of 2 m; the jump efficiency is the ratio of the height in air (h_a) and in a vacuum (h_v) (from Bennet–Clark and Alder, 1979).

Animal	Mass	Height in air (h_a)	Jump efficiency (h_a/h_v)
Flea	0.3 mg	0.53 m	0.27
Locust	440 mg	1.48 m	0.74
Galago	200 g	1.93 m	0.97

a galago has jumped to 2.25 m (Hall–Craggs, 1965). It is unlikely that its muscles can produce much more force per square centimeter cross section than those of other animals, and the superior performance can be explained only by a larger muscle mass involved in the jump (more energy for the takeoff contraction) and possibly by a more favorable mechanical structure of the limbs.

The galago does have large jumping muscles, nearly 10% of the body mass (Alexander, 1968), or about twice as much as in a man. If all combined mechanical advantages of this highly specialized jumping animal could account for a 50% increase in performance, this, combined with a doubling in the muscle mass, could explain the threefold increase in jumping performance. As a standing jump, this is more impressive than that of the flea, but recall that the galago is a highly adapted warm-blooded animal, making long jumps in its natural jungle habitat.

Elastic energy storage

There is a constraint on jumping that can be resolved only by a radical change in the design of the jumping mechanism for small animals. The reason is simple: The smaller the animal, the shorter its takeoff distance (Table 14.2). To reach the necessary takeoff speed, which theoretically should be the same as for a large animal, the small animal must accelerate its body much faster. The time available for the takeoff is very short (Table 14.2), and muscles just cannot contract that fast.

For a flea, acceleration takes place over less than 1 mm, and takeoff time is less than 1 msec. The average acceleration during takeoff must therefore exceed 200 g. It is worth a moment's reflection to think of what such high acceleration means. It means that the force on the animal is 200 times its weight (any mammal would be totally crushed under such forces), and the insect must have a skeleton and internal organs able to resist such acceleration forces.

If muscle contraction were used directly to accelerate a flea, the flea could not jump at all. The problem is solved by using the principle of the catapult. Energy is stored in a piece of elastic material, resilin, at the base of the hindlegs. The muscles are used to compress the resilin, which is a rubber-like material that returns the energy with close to 100% efficiency when a release mechanism is tripped. The elastic recoil works much the same way as a slingshot and imparts the necessary high acceleration to the flea. This principle of storing energy in an elastic system is also used by other small jumping animals, for example, the click beetle. Larger animals, such as mammals, use the muscle contraction directly to impart the necessary acceleration, because their takeoff times are sufficiently long.

The analysis of the jumping mechanisms of various animals provides an excellent example of the merits and limitations of scaling considerations. Simple considerations tell us that all animals, irrespective of size, should be able to jump to approximately the same height. For the smallest animals, air resistance is a limitation, because their surface areas, relative to mass, become increasingly important. Careless use of scaling considerations could, however, overlook the design limitations on the very small animal. Without a redesign of the jumping mechanism, in this case by providing force through elastic storage, there would be a constraint that would make it impossible for a flea to jump.

15

Swimming and flying

Animals running on land are supported by a solid substratum. Animals that swim and fly move in fluid media and have no solid support; they are supported by the medium through which they move. Fish have nearly the same density as water, and the energy they use for locomotion goes into overcoming the resistance of the medium. A flying bird must also overcome the resistance of the medium, but in addition it must keep from falling to the ground; that is, it must provide lift equal to its body weight.

Fish

Because fish are nearly neutrally buoyant, they expend little or no energy to support themselves, but energy is needed to overcome the resistance of the medium. The resistance that a swimming fish encounters is called the drag. To overcome the drag, the fish must provide thrust that equals the drag. There are two components to the drag on a fish moving through the water: pressure drag and friction drag.

Friction drag can be thought of as the drag on a thin, flat plate being pulled through a fluid parallel to its plane. *Pressure drag* can be thought of as the drag on the plate if it is moved through the fluid in a direction vertical to its plane.

Pressure drag is difficult to calculate accurately. It comes from the necessity to displace water during forward movement, and it is determined by the frontal area (the projected body area onto a plane normal to the direction of swimming) and by the shape of the body. A well-designed, streamlined body, such as a fish, has a much lower pressure drag than, say, a sphere.

The friction drag corresponds to the force used to overcome the viscosity of the fluid. The fluid immediately in contact with the surface of a moving body is carried along with it, and this establishes a layer of shear (a velocity gradient) in a thin layer adjacent to the moving body (the boundary layer). The force needed to overcome the viscosity is known as the friction drag.

The friction drag, D_f, can be calculated from the equation

$$D_f = \tfrac{1}{2}\rho S U^2 C_f$$

where ρ is the density of the fluid, S is the surface area, U is the speed, and C_f is the drag coefficient.

This equation is the basis for the common statement that the drag increases with the square of the velocity, which implies that the drag coefficient, C_f, is a constant. This is not so, because C_f varies with the Reynolds number, which we therefore must introduce. The Reynolds number is a nondimensional number that represents the ratio between inertial and viscous forces. For a laminar boundary layer, C_f varies inversely with the square root of the Reynolds number, Re, which in turn varies directly with the speed, U. Other variables remaining constant, the drag coefficient decreases with increasing Reynolds number (and speed), up to a limit. If the boundary layer at higher speeds is turbulent, the drag coefficient also varies, in this case inversely with the fifth root of the Reynolds number.

The complex functions that determine pressure drag and friction drag make it extremely difficult to evaluate on a theoretical basis the power requirements for fish of different sizes swimming at differing speeds. For a well-designed, streamlined fish, the total drag coefficient, C_D, will be of the order of $1.2\,C_f$ (Webb, 1978).

What does the drag cost the animal in terms of energy? What we like to know, in fact, is the energy per unit time, which is the same as metabolic power or metabolic rate. Power equals force (in this case thrust) multiplied by speed. Drag is a force and has the dimensions $M\,L\,T^{-2}$. Speed has the dimensions $L\,T^{-1}$. The product of drag and speed therefore has the dimensions $M\,L^2\,T^{-3}$, which are the dimensions of power (work per unit time), that is, metabolic rate or metabolic power.

We found earlier that drag is proportional to speed squared (U^2). Multiplying by speed gives the metabolic power that goes into swimming, which therefore increases with the cube of speed (U^3).

This is the simple physics of it; next, let us turn to real fish.

Table 15.1. Rates of oxygen consumption ($\dot{V}_{O_2}$, in mg O_2 per hour) related to body size (M_b, in g) for salmon swimming at various speeds; the maximum sustained speed is the speed that could be maintained by a fish for 60 min; the standard rate approaches the resting metabolic rate; all determinations were at 15°C (Brett, 1965).

	Rate of oxygen consumption
Standard $\dot{V}_{O_2}$	$\dot{V}_{O_2} = 49.3\, M_b^{0.775}$
1/4 maximum speed	$\dot{V}_{O_2} = 103.5\, M_b^{0.846}$
1/2 maximum speed	$\dot{V}_{O_2} = 205.6\, M_b^{0.890}$
3/4 maximum speed	$\dot{V}_{O_2} = 358.9\, M_b^{0.926}$
Maximum sustained speed	$\dot{V}_{O_2} = 724.6\, M_b^{0.970}$

Swimming salmon

The most comprehensive studies of the energetics of swimming fish have been accumulated by the Canadian investigator J. R. Brett. He measured the oxygen consumption in salmon ranging from 3 to 1400 g in body size over a wide range of speeds. Larger salmon were available but would not fit into his apparatus, but even so, his data cover nearly three orders of magnitude.

It is no surprise that large salmon can swim faster than small ones. The relationship given by Brett (1965) gives the swimming speed, U (cm/sec), in relation to the length of the fish, l (cm), as

$$U = 19.5\, l^{0.50} = k M_b^{0.17}$$

In other words, the swimming speed for salmon increases with the square root of the length of the fish and the 0.17 power of body mass. This precise relationship is interesting in view of what we shall learn about flying birds.

Brett measured oxygen consumption of salmon at rest and when swimming at various speeds. The maximum swimming speed that a fish could sustain for 1 hr was established, and the rate of oxygen consumption was determined at this speed, and also at a series of lower speeds: 3/4, 1/2, and 1/4 of the maximum sustained speed. For reference, the standard or resting metabolism was also determined.

The equations that relate the rates of oxygen consumption at these various swimming speeds to the body mass are listed in Table 15.1. The standard rate of oxygen consumption over the 500-fold size range varied

Figure 15.1. Relationship between oxygen consumption and body mass for salmon swimming at different speeds. The maximum speed is the speed the fish could sustain for 60 min, and lower speeds are expressed as fractions of the maximum speed. From Brett (1965).

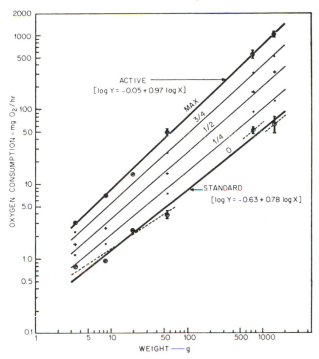

with the exponent 0.775. This exponent is similar to many other determinations on fish, which often cluster around the value 0.8. The rate of oxygen consumption at the maximum sustained speed, on the other hand, increased almost exactly in proportion to the body mass (the exponent 0.97 ± 0.05 is within the 95% confidence limits of 1.0).

Because the rate of oxygen consumption at maximum activity increases with body size much faster than the standard rate of oxygen consumption, the larger fish must have a much greater factorial metabolic scope than small fish. The ratio between the highest sustainable rate relative to the standard rate (the factorial scope) was 4.0 for the smallest fish. For the largest salmon this ratio increased to 16; that is, the sustainable rate was 16 times as high as the standard rate.

These relationships are also shown in Figure 15.1 (Brett, 1965). At first glance, the slopes of the regression lines may not appear very different, mainly because the ordinate is on a logarithmic scale. However,

Figure 15.2. Cost of swimming for salmon of various body sizes. The equation for the regression line is cost $= 0.426\,M_b^{-0.254 \pm 0.017\,\text{SE}}$ (where cost is in ml O_2 g^{-1} km^{-1} and M_b is in g). Calculated from data reported by Brett (1965). See Table 14.1 for comparison with land animals running.

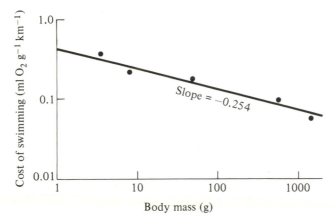

the factorial scope increases from about 4 for the smallest fish to about 16 for the largest. The change in slope that to our eye appears fairly small represents a substantial difference over the 500-fold size range that we are concerned with in this graph.

Can the data collected by Brett be used to calculate a cost of locomotion in the way that we did for running mammals? The cost of running was linearly related to the speed of running, and therefore the slope of the regression line could be used directly as a measure of the cost of locomotion.

The oxygen consumption of swimming fish is not linearly related to the swimming speed, and we must therefore choose some other way of analyzing the vast amount of information obtained by Brett. Perhaps it is arbitrary to choose the swimming speed that is 3/4 of the maximum speed that the salmon can maintain for 1 hr, but it will give us some indication of how the cost changes with body size. By dividing the rate of oxygen consumption (ml O_2 g^{-1} hr^{-1}) by swimming speed (km hr^{-1}), we obtain the cost of swimming in the units we used before for running mammals (ml O_2 g^{-1} km^{-1}). The data are plotted in Figure 15.2, and the regression line has a slope of approximately -0.25. Choosing other swimming speeds for the calculation would give different lines, but all would have negative slopes, and of about the same magnitude, showing that it is cheaper for a large fish to transport 1 g of body mass over a given distance.

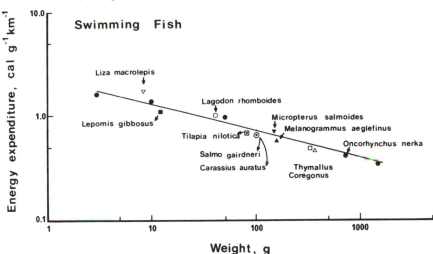

Figure 15.3. Energy cost of swimming relative to body size for a variety of fish, as reported by Beamish (1978).

The data in Figure 15.2 pertain only to salmon, a species in which the small and the large fish have similar shapes. It is interesting, however, that the data for many other species of fish fall on approximately the same regression line as the salmon (Figure 15.3). Here the slope of the regression line is approximately -0.3, not too different from that for the entire range of salmon.

It is peculiar that not only fish that are similar in shape to salmon but also others (such as pinfish and tilapia) fit the curve for salmon, although their shapes are quite different from that of the fast-swimming salmon. Even the data for eel tend to fall in the same range, although their mode of swimming is entirely different: They use the entire body in propelling themselves through the water (anguilliform propulsion), whereas other fish use the tail portion of the body (carangiiform propulsion). In spite of these differences in shape and swimming mode, the energy cost of propulsion appears to be similar when calculated for moving one unit of body mass over one unit of distance.

The next problem will be to compare this finding with data for the other large group of animals that move in a fluid medium, the birds.

Flying animals

Some eight or nine thousand bird species are known to science. A few birds are flightless, but most are excellent flyers. The size range for flying birds is from about 3 g to 10 kg, more than three orders of magnitude, and the largest birds are all flightless. Only birds of small body size are able to hover in one place, but nearly all can move relatively fast in forward flight. Only the smallest, especially the hummingbirds, are able to hover for any length of time.

Of the more than 1 million animal species described, at least three-quarters are winged insects, and most of these can practice hovering or slow forward flight, rather than fast forward flight. Their sizes span six or more orders of magnitude, from less than 25 μg to more than 25 g. This is as wide a size range as we find for land mammals, which also span six orders of magnitude (3 g to 3 000 000 g).

Birds

There is an immense amount of information available about birds and their morphological features. However, when it comes to physiological information, particularly the energetics of flight, our information is very inadequate. One main reason is that studies of birds during flight are technically very difficult; such studies are usually performed in wind tunnels and require long training periods for the birds and difficult instrumentation for the investigator. Good information about the energetics of bird flight is available only for a handful of species, ranging from about 30 g to 300 g in size. (We have information about hovering flight of hummingbirds, but the energetic requirements for hovering differ substantially from fast forward flight.)

The meager information on birds is far less satisfactory than what we know about the many species of mammals that have been studied. Our understanding of bird flight is therefore to a great extent based on aerodynamic theory and the highly advanced technology of airplanes and helicopters.

The structure of birds

A tremendous amount of information about the flying apparatus of birds has been assembled and evaluated in an impressive publication by Greenewalt (1962). The following equations represent some of the most important information from Greenewalt's data:

wing length $\propto M_b^{0.33}$

wing area $\propto M_b^{0.67}$

pectoral muscle, large $= 0.155 M_b^{1.0}$

pectoral muscle, small $= 0.016 M_b^{1.0}$

Although there is a considerable amount of individual variability, on the whole these numbers tell us that birds, as a group, are surprisingly close to being isometric. Wing length and wing area are scaled directly in proportion to l and l^2, respectively. (Hummingbirds seem to differ in a systematic way from other birds, because their wing length is closer to being proportional to $M_b^{0.67}$, undoubtedly related to hovering as the primary characteristic of their flight.)

The power plant for flight (the flight muscles) is another interesting matter. The large pectoral muscles average 15.5% of the body mass, irrespective of the size of the bird, and the small pectorals approximately one-tenth as much. It is remarkable that this proportion remains the same from a tiny kinglet to the mute swan. Together the flight muscles average 17% of the body mass of ordinary birds. The dominating size of the large pectorals suggests that the power for flight is provided wholly by the downstroke of the wings. There are some systematic deviations; for example, birds of prey usually have even smaller small pectorals.

For hummingbirds, the situation is different. The large and small pectorals combined account for 25 to 30% of the body mass, with the ratio of large to small pectorals being roughly 2:1, as compared with 10:1 for ordinary birds. The larger size of the small pectorals suggests that power for flight is contributed both during upstroke and downstroke of the wings, which indeed has been confirmed by analysis of high-speed films.

Flight speed

Observed flight speeds for birds are of little value if related to a point on the ground (the ground speed), because the velocity of the air at the height of the bird usually is unknown. Much of the information in the literature is therefore of no use. Observations obtained by accurate triangulation, with corrections for measured air speeds, were obtained by Tucker and Schmidt–Koenig (1971). They observed a twofold range in speed, from 10.8 to 20.7 m/sec, for 22 species of birds. This twofold range was not obviously related to the sizes of the birds, which ranged from red-winged blackbird to Canada geese, a size range of two orders of magnitude. However, the flight patterns were very different; for example, ducks (fast flyers) and herons (slow flyers) move their wings very differently. Such differences are likely to mask any systematic body-size relationship that may exist.

Scaling arguments developed by Lighthill (1974) on the basis of aerodynamic theory suggest that the characteristic speed for flying birds should vary as $l^{0.5}$ or $M_b^{0.17}$. Over a 100-fold difference in body sizes, this should give slightly more than a twofold difference in cruising speed, but a twofold difference is probably less than the noise in the system caused by the different flight patterns that are characteristic of different types of birds.

Greenewalt (1975*b*) arrived at a relationship similar to the theoretical result. His data suggested that the flight speed should be proportional to the wing loading to the power 0.55. Wing loading (weight per unit wing area) in turn is proportional to *l*, and the flight speed for birds should therefore be

$$U \propto l^{0.55} \propto M_b^{0.18}$$

This is virtually identical with the relationship suggested by aerodynamic theory.

Tucker, in his extensive theoretical treatment of bird flight (1973), included the expected flight speeds at which cost of transport is at a minimum, as given by

$$U = 14.6 M_b^{0.20}$$

where U is in meters per second and M_b is in kilograms. The proportionality factor (14.6) will be lower (13.1) if the wingspan is 20% greater, and higher (16.7) if the wingspan is 20% shorter. For a 100-g bird, the average equation will give a flight speed of 9.2 m/sec, and for a 3-kg bird, 18.2 m/sec, a twofold range, which is roughly the range observed in the measurements of air speeds made by Tucker and Schmidt–Koenig (1971).

The good agreement between birds and scientists makes it appear that birds know a great deal about aerodynamic theory.

Drag and cost of flight

Birds, like fish, move in a fluid medium and experience the same kinds of drag: friction drag and pressure drag. However, whereas fish are supported by the medium, birds must constantly provide lift in order to stay aloft and keep from falling down. For horizontal flight, the lift must equal their weight. The need to provide lift is experienced as drag, known as *induced drag*. The lift of a wing is proportional to the density of the air (ρ), the area of the wing (S), approximately to the square of the velocity through the air (U), and to a lift coefficient (C_L):

$$L = \tfrac{1}{2} \rho S U^2 C_L$$

Both lift and drag vary with the shape of the wing, and especially with the angle of attack, but in very different ways. Lift increases with U^2, but the induced drag is inversely proportional to U^2, which makes flight at low speeds very expensive. These relationships make it extremely difficult to predict accurately the expected cost of flight for a bird.

A theory for calculating the relationship between power required for flight and body size has been developed by Pennycuick (1969), based on aerodynamic helicopter theory. This work was further developed by Tucker (1973), who incorporated in the theory the available observations on the power input of birds as measured in wind-tunnel experiments. The birds ranged from budgerigar (35 g) to laughing gull (322 g), and included fish crow (275 g), as well as two species of bats, the largest, the flying fox, weighing 780 g.

In his careful analysis of bird flight, Tucker developed simplified equations for estimating power input (metabolic rate) as being directly proportional to body mass:

$$P_i = 84.7 \, M_b^{1.0}$$

where P_i is power in watts and M_b is body mass in kilograms.

Maximum size for bird flight

If the metabolic power required for flight increases in proportion to body mass, and the available metabolic power increases at some lower rate, there will be a maximum body size above which active flapping flight will not be possible. One of the largest birds that can fly is the Kori bustard (*Ardeotis kori*), which weighs about 13 kg. It is a ground-living bird that seldom flies, and then only briefly. Vultures and albatrosses that weigh nearly as much fly for long periods, but much of their flight consists of soaring in wind or thermals.

There have been no direct determinations of the metabolic power of large flying birds, and measurements of their resting metabolic rates are of little help, because we do not know the metabolic scope for activity. In the absence of better information, the preceding argument can be presented in a simple graph (Figure 15.4).

The argument pertaining to a maximum body size for flight was used by Wilkie (1959) to evaluate whether or not man-powered flight is possible. Wilkie reached the conclusion that a well-trained human athlete should just be able to keep a very light plane aloft. His conclusion was later vindicated by the successful flight of a man-powered plane, the Gossamer Albatross, which was designed with the utmost consideration of every possible advantage and in 1979 was able to cross the channel from England to France.

At first thought, it may seem unlikely that insects would similarly be constrained as to a maximal size compatible with the power needed for flight. However, in studies of hovering moths, Casey (1981) found that

Figure 15.4. Power required for bird flight compared with available power. A line for resting metabolic power is drawn with a slope of 0.72. A second line with the same slope is drawn at 10 times the resting level (the factorial scope may in fact be higher). A third line with the slope of 1.0 represents the estimated power required for flight. This line is arbitrarily drawn to intersect with the activity line at 10 kg body mass, the approximate size of the largest flying birds. The lines show that the power required for flight increases more rapidly than available power, setting a limit to the maximum size for a flying bird.

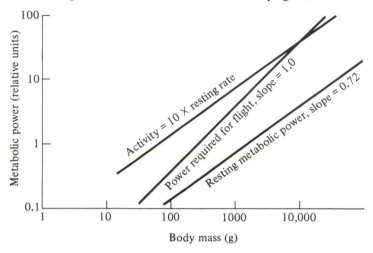

the aerodynamic power requirement increases with size more rapidly than available metabolic power. Aerodynamic power for hovering increased with $M_b^{1.08}$, whereas the available metabolic power increased with $M_b^{0.77}$, as shown in Figure 15.5.

The two lines in Figure 15.5 would, if extended, intersect at a body mass of about 100 g. One wonders if it is a coincidence that 100 g is also the largest body size for birds that are able to hover for more than a few moments.

Hovering flight is metabolically three times as expensive as forward flapping flight. The maximum size for a forward-flying bird should therefore be 81 times the maximum size for a hovering bird. This is arrived at in the following way. If the specific metabolic power decreases with $M_b^{-0.25}$, it takes an 81-fold increase in size to decrease the specific power threefold. Relative to 100 g as the maximum size for hovering, the limit for forward flight should be about 8 kg, not far from reality. Again, this very approximate calculation, which in fact is not too well supported in theory, merely points to an area that is in need of more experimental work.

Figure 15.5. Metabolic power output for hovering sphinx moths relative to body mass (upper curve) and the aerodynamic power required for hovering (lower curve). If the two regression lines are extended, they will intersect at a body mass of 100 g. From Casey (1981).

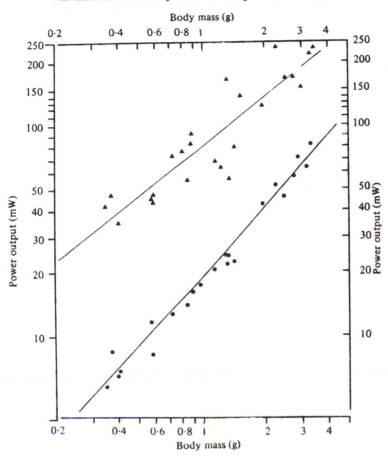

Is there a lower size limit?

The upper size limit for flying birds illustrates convincingly that there are constraints on scaling that preclude unlimited increases in size. What about small size? Are there other constraints that form a lower size limit for flight?

We know that the smallest birds are hummingbirds, which weigh between 2 and 3 g. They not only are very fast in forward flight but also are able to hover for long periods of time, apparently with the greatest ease. Hummingbirds may represent the smallest possible size for a bird, but by no means for flight in general. Insects that weigh as little as a small fraction of a milligram are perfectly able to fly, but there are limitations that require novel mechanisms for flight.

The generation of lift by classic motion of an airfoil becomes less and less effective at smaller Reynolds numbers. The Reynolds number decreases with decreasing speed and with decreasing linear dimensions, and viscous forces then dominate over inertial forces. The wings of a tiny fruit fly are surrounded by a thick boundary layer of air, which is the cause of very poor lift–drag ratios. The drag increases, and the attainable lift is increasingly inadequate.

This situation represents a constraint on small size. Decreasing the size below certain limits brings us into a world in which the physical laws appear different, and entirely different mechanisms must then be employed to achieve lift. Weis-Fogh (1973) studied a tiny wasp, *Encarsia*, which uses a method for achieving lift that has been designated the clap-and-fling mechanism. This tiny insect weighs no more than 0.03 mg and hovers with its body in the vertical position. It moves its wings about 400 times per second, beginning with the wings clapped together behind its back. As the wings are suddenly flung open and rotated, air rushes in. This fling motion generates a circulating vortex of air about each wing that produces the necessary lift.

At a body mass of 0.03 mg, *Encarsia* is five orders of magnitude smaller than the smallest hummingbirds. Evidently, for reasons concerning the aerodynamics of flight, a smaller hummingbird would be feasible, but there may be other constraints that set a lower limit to the size of hummingbirds. The methods of oxygen supply are radically different in birds and insects, and this may be where we should seek an explanation. We shall return to this question in the next chapter, when we discuss the lower size limit of about 3 g for warm-blooded mammals and birds, the shrews and the hummingbirds. We shall compare these with insects that can both keep warm and fly.

Cost of transport

Can we meaningfully compare running and flying? It is not easy, but in spite of the fundamental differences, we can arrive at some useful conclusions.

We saw earlier that oxygen consumption during running increases regularly and linearly with increasing speed. For birds, the situation is very different. Birds usually have an optimum speed of flight at which their oxygen consumption is at a minimum. Flying slower or faster increases

the metabolic rate; at very low speeds the induced drag increases tremendously, and at higher speeds the power to overcome friction and
pressure drag increases with the third power of the speed.

The first broad investigation of oxygen consumption for a bird flying
at a variety of different speeds, as well as on up- and down-sloping
paths, was made by Tucker (1968) in an extraordinarily important study.
Because of the difficulties involved, other similarly complete investiga-
tions have been made on only a small number of bird species. Similar
studies on bats by Thomas (1975) have added important information.

The energy cost of flight appears to be similar for birds and bats, and
in pulling together the available data for the minimum cost of transport
for each species, ranging in size from Tucker's budgerigar (35 g) to a
flying fox (779 g), Thomas gave the following equation:

$$P^* \propto M_b^{-0.21}$$

where P^* is the specific power (specific metabolic rate) and M_b is body
mass.

The cost of transport for a flying bird or bat can be calculated as follows: If we take the metabolic rate during flight at the most economical
speed and divide this by the flying speed, we obtain a cost figure in the
same units we used for mammals: milliliters of oxygen used to transport one unit of body weight over one unit of distance. The difference
between birds and mammals is that the cost will differ for a bird flying at
any other speed, whereas the cost for mammals (because of the straight-
line relationship) is independent of speed. Another difference is that in
the numbers used for the bird calculation, the nonflying fraction of the
metabolic rate (the resting rate) is included. The numerical difference will
not be very great, because flight metabolism is roughly 10 times the resting metabolism; that is, including the resting rate introduces only a
minor error.

Returning to the earlier discussion, the specific metabolic rate (specific
power) during flight should, according to aerodynamic theory, increase
with $M_b^{0.17}$. However, according to the analyses by Tucker, in which
actual determinations of flight metabolism were included, the specific
power during flight should be independent of body size (proportional to
M_b^0). Let us for the moment use Tucker's relationship.

The speed of flight for birds varies a great deal. According to Greenewalt's compilations (1975b), the characteristic speed for flying birds
should vary with $M_b^{0.18}$, with slight variations from one group to another
depending on their flight characteristics. Pennycuick (1969) used $M_b^{0.17}$ in

his analysis of bird migration. If we take Tucker's power for flight and divide by the speed, the cost of transport should decrease with increasing size according to

$$\frac{\text{metabolic power}}{\text{speed}} = \text{cost} \propto M_b^{-0.17}$$

Greenewalt estimated the energy required per unit weight per unit distance and found that the cost decreased with increasing body size, the slope of the regression line being -0.15 (Greenewalt, 1975b). We do not have sufficient accurate information about metabolic rates for birds flying at normal speeds to evaluate these relationships. The handful of birds that have been studied have quite different flight patterns (e.g., crow and laughing gull), and we need more information to evaluate the scaling effect of body size.

If insects are included, we can cover a much wider size range, from less than 1 mg to over 100 g, more than five orders of magnitude.

In a compilation that included insects, Tucker arrived at the following equation for the cost of flying:

$$\text{cost (liter } O_2 \text{ kg}^{-1} \text{ km}^{-1}) = 0.26 \, M_b^{-0.23}$$

This amply confirms that the specific cost of transport for flying decreases with increasing body size, as it does for animals running on land.

For any given bird, the cost of transport is much less than the cost for a running animal of the same size. This is easily explained. The bird flies much faster than an animal of the same size can run. Mammals and birds have roughly the same resting metabolic rates, and for both the factorial scope for activity is approximately one order of magnitude higher. If a bird, for a given power output, moves much faster than a running mammal, it covers a greater distance for the same metabolic cost; hence, the cost of transportation is proportionately lower for the bird.

What we have seen are several similarities in the scaling of swimming and flying relative to body size. This may seem surprising in view of the different physical nature of water and air and the conspicuous differences in the structure of fish, birds, and insects.

16

Body temperature and temperature regulation

Birds and mammals maintain their body temperatures more or less constant and independent of variations in environmental temperature. To maintain a constant body temperature, there is one fundamental requirement: The rate of heat loss must equal the rate of heat production. This simple fact is basic to all considerations of temperature regulation.[1]

Both heat production and heat loss can be varied through a number of physiological mechanisms. Most mammals and birds are very successful in achieving balance and maintaining a constant core temperature with only minor fluctuations. One regular fluctuation is the change in core temperature with the day-and-night cycle, the night temperature being a couple of degrees lower than the day temperature in diurnal animals, and vice versa for nocturnal animals. The body temperature is also increased during physical activity, but we shall not be concerned with any of these fluctuations.

What do we mean by body temperature? This is by no means a simple question, because the different parts of the organism are not all at the same temperature. Body temperature in mammals and birds is usually taken to mean the temperature of the deeper abdominal organs, sometimes referred to as the *core temperature,* and often measured as the deep rectal temperature.

Scaling of heat loss

We saw earlier that various groups of warm-blooded vertebrates have characteristic ranges of body core temperatures, without any

---

1 Some metabolic energy may go into the performance of external work, but this does not enter into the heat-balance equation of the organism. However, if the calculation of heat production is based on oxygen consumption, any external work must be known and subtracted.

Table 16.1. Approximate ranges of body core temperature for major groups of mammals and birds; within each group, the core temperature is independent of body size, i.e., scale-independent.

Animal group	Core temperature
Monotremes (echidna, platypus)	30–31°C
Marsupial mammals	34–36°C
Eutherian mammals	36–38°C
Birds	39–41°C

obvious relationship to body size (Table 16.1). In each group, the body temperature is maintained within a few degrees as a scale-independent variable.

The absence of any obvious relationship between body temperature and body size and the clear scaling of heat production (metabolic rate) in relation to size (see Chapter 6) make it necessary to consider the scaling of heat loss. Obviously, heat loss must be scale-dependent, because heat production is scale-dependent. Also, it is intuitively obvious that if a small and a large animal maintain the same body temperature, more heat is lost from the larger animal.

When it comes to heat loss, we shall operate with a highly simplified concept: the thermal conductance of the animal. This concept will take the place of a full account of the extremely complex situation of heat loss from the living animal. Heat produced in the core is transported to the surface partly by conduction through the tissues, but mostly by convection (circulation of blood). From the skin surface, heat flows by conduction through fur or feathers to the environment, where conditions such as air temperature, air convection, and radiation enter into the heat-exchange processes as highly complex functions.

Let us turn to the definition of overall conductance (C), which simply says that the rate of heat flow ($\dot{H}$) is proportional to the temperature difference between the body (T_B) and the environment (T_A):

$$\dot{H} = C\,(T_B - T_A)$$

We can immediately see that, if heat production and body temperature are constant, C must change if T_A changes. An increase in T_A, that is, a smaller difference between air and body temperature, requires an increased C to maintain a constant rate of heat flow. Conversely, a lower air temperature (increased $T_B - T_A$) requires a lower conductance to maintain balance. The conductance can be changed only within certain limits, primarily by changes in blood flow to the surface and by increasing the exposed surface by stretching out the limbs and exposing thinly furred skin areas.

Below a certain point, conductance cannot be further decreased. Think of conductance as the reciprocal of insulation. Conductive heat loss can be decreased by increasing the insulation. Decreasing the blood flow to the skin reduces heat flow from the core to the skin surface; fluffing up fur or feathers increases the external insulation; curling up reduces the exposed surface and heat loss. However, there are limits to these countermeasures, and when the air temperature falls below a certain limit (the lower critical temperature), the animal can maintain its body temperature only by increasing heat production. At the lower critical temperature (LCT), the conductance has been minimized as far as possible, and at lower temperatures heat production must be increased. We ourselves are familiar with this situation, because we shiver in the cold, which is our way to increase heat production when insulation is insufficient.[2]

The minimal conductance, that is, the conductance at or below the LCT, must vary with animal size. There are two obvious reasons: One is that larger animals have smaller relative surfaces. The other is that larger animals usually have thicker fur than small animals. This means that large animals are favored because of their smaller relative surface as well as better surface insulation. (For the moment we shall disregard the largest tropical mammals, elephant, rhinoceros, and some others that are hairless.)

When we compare conductance and heat production (metabolic rate), it will be convenient to express both in relation to body size. Thus, we shall compare specific conductance (C^*, heat flow per unit body mass) with specific metabolic rate ($\dot{H}^*$). We shall be concerned only with minimal thermal conductance, which in the following will be referred to simply as conductance.

A considerable amount of material on thermal conductance in birds and mammals was accumulated and discussed by Herreid and Kessel

---

2 This discussion disregards evaporation of water, which at high air temperatures is an important avenue of heat loss. At low temperature, evaporation represents a small fraction of the total heat loss, and we shall not account for it separately.

Table 16.2. Thermal conductance in relation to body size for mammals and birds, given by the equation $C^* = aM_b^b$, where C^* is specific thermal conductance [ml O_2 (kg hr °C)$^{-1}$] and M_b is body mass (kg).

Animal group	Number of species	Size range (kg)	a	b
Eutherian mammals[a]	24	0.003–0.598	31.2	−0.505
Eutherian mammals[b]	24	0.003–0.598	30.2	−0.52
Eutherian mammals[c]	192	0.0035–150.0	40.1	−0.426
Marsupials[b]	12	0.0072–5.050	37.3	−0.463
Birds[a]	31	0.0106–2.755	23.6	−0.536

[a] Herreid and Kessel (1967).   [b] MacMillen and Nelson (1969).
[c] Bradley and Deavers (1980).

(1967). This was followed by several more recent compilations on thermal conductance relative to body size. Some of the resulting equations are listed in Table 16.2. All these equations have been recalculated to the same units, expressing conductance as milliliters of oxygen per kilogram of body weight per hour per degree Celsius temperature difference. The reason for using oxygen is that all the data originally were based on determinations of the rate of oxygen consumption for these animals. The coefficient can readily be recalculated to other units by equating 1 ml of oxygen with 4.8 cal, or 1 liter of oxygen per hour with 5.579 W.

Are there any meaningful differences between these equations? Let us look at the first three: those for eutherian mammals. The mass exponents (b) differ somewhat, and so do the coefficients. However, the regression lines they represent intercept, and for a body size of 0.1 kg, the equations give virtually identical conduction values. Consider that most of the data points are clustered around this size range, and furthermore, that the samples represent different populations. We must conclude that there probably is no biological difference that we can extract from the information at hand. The data are compiled from a wide variety of sources, and stated simply, even a statistically significant difference in this case probably would be biologically meaningless.

Marsupials do not differ appreciably from eutherian mammals, but when it comes to birds, these seem to have a significantly lower conductance. In the original equations given by Herreid and Kessel (1967), the proportionality coefficients for mammals and birds did not differ, but their equations expressed the body mass in grams, that is, a unit mass

Table 16.3. Thermal conductance for mammals and birds, measured at rest, but during their normal activity period (the α period) and normal rest period (the ρ period); units as in Table 16.2 (Aschoff, 1981).

Animal group	Number of species	Size range (kg)	a	b
Eutherian mammals α	27	0.079–6.660	43.3	−0.517
Eutherian mammals ρ	59	0.004–4.440	28.3	−0.519
Birds, passerine α	28	0.006–1.130	35.0	−0.463
Birds, passerine ρ	26	0.0103–0.360	23.8	−0.461
Birds, nonpasserine α	39	0.0027–2.430	33.4	−0.484
Birds, nonpasserine ρ	11	0.040–2.020	16.9	−0.583

entirely outside the range for any adult bird or mammal. What we have is two regression lines that converge at about 1 g body size, and at larger body size the line for birds is significantly below the mammalian line.

The concept of conductance was critically evaluated by Aschoff (1981), who discussed the meaning of the term "conductance" and evaluated a large mass of observations for both mammals and birds. Aschoff found, as earlier authors had before him, that specific heat conductance varies with $M_b^{-0.5}$ (Table 16.3). However, he emphasized that there was a conspicuous difference between measurements obtained during the normal activity period and the normal rest period for the same animal. Although all determinations were made with animals at rest, diurnal animals had consistently higher metabolic rates in the daytime (their normal activity or α period) than at night (their normal rest or ρ period). Conversely, determinations for nocturnal animals were higher at night (the α period for these animals) and lower during the day (the ρ period for nocturnal animals). It should be emphasized that all animals were at rest, whether measurements were made during the α or the ρ period.

The role of fur

Large animals tend to have thicker fur than small animals. A mouse or a lemming simply could not move about if it had fur as thick as a fox or a bear. Heavy fur has a lower conductance, and large animals are better protected against heat loss. Can these differences be directly related to the decrease in conductance with increasing body size?

The heat loss from an animal takes place mainly from the outer surface, and larger animals have relatively smaller surface areas. Let us examine how this relates to the decrease in conductance with increasing body size. We have

specific heat conductance, $C^* \propto M_b^{-0.50}$

relative body surface area, $S^* \propto M_b^{-0.33}$

The mass exponents in these two expressions show that conductance decreases more rapidly with increasing body size than does body surface area; in other words, insulation is greater in the large animal. Should we make the careless suggestion that the change in conductance resides entirely in the increased fur thickness, the conclusion would be as follows. By dividing the two equations, we find that the conductance per unit area is proportional to $M_b^{-0.17}$. That is, conductance per unit area decreases with the mass exponent -0.17, or insulation increases with the mass exponent 0.17.

The information we have about the thickness and insulation value of fur is insufficient for a precise evaluation of its scaling relative to body size. The data we have indicate that the insulation value increases with body size up to about 10 kg, and among larger animals there are variations without any clear relationship to body size. This trend emerged from Scholander's study of fur from 15 mammals (1950b). A later study of 10 mammals, ranging from a 20-g deer mouse to a 430-kg polar bear, indicated the same trend: Up to 10 kg there was a fairly regular increase in the insulation value of the fur, and above that limit there was no clear relationship to size (Hart, 1956). The slopes for regression of the insulation value of the fur on body size, using the data for sizes less than 10 kg, yield mass exponents between 0.15 and 0.20 for both sets of data. In Hart's study, winter fur had consistently higher insulation value than summer fur from the same animal, but the body-size relationship remained the same with a mass exponent of about 0.20 in each case.

It would be rash to conclude that the decrease in conductance with increasing body size is due entirely to the thickness of the fur. However, for relatively small animals, less than 10 kg or so, the thickness of the fur appears to be an important factor. Other considerations may also be important: perhaps more than anything else the thickness of subcutaneous fat, which in large animals can be substantial.

Conductance and tolerance to cold

What is the consequence of the lower conductance in the large animal? A comparison of conductance and heat production is revealing, because conductance decreases with body size more rapidly than does specific heat production. This is expressed by the mass exponents in the following relationships:

$$\text{specific heat conductance, } C^* \propto M_b^{-0.50}$$

$$\text{specific metabolic rate, } \dot{H}^* \propto M_b^{-0.25}$$

The fact that larger animals are better protected against heat loss simply means that a larger animal will have a lower critical temperature below which heat production must be increased in order to remain in heat balance. Let us return to the definition of thermal conductance:

$$\dot{H}^* = C^* (T_B - T_A) = C^* \Delta T$$

We therefore have

$$\Delta T = \frac{\dot{H}^*}{C^*} \propto \frac{M_b^{-0.25}}{M_b^{-0.50}} \propto M_b^{0.25}$$

This expression states that because, with increasing body size, specific thermal conductance (C^*) decreases faster than specific heat production ($\dot{H}^*$), the ΔT that the animal can support without increasing heat production must increase with size with a mass exponent of 0.25. The result is simple: Large animals have lower critical temperatures than small animals, which has been well known since the classic work of Scholander and associates (1950*a*).

There is, however, a notable consequence of the mismatch between heat production and conductance as body size is increased. The larger the animal, the more difficult it will be to eliminate the metabolic heat. Even if we were to remove all fur, a very large mammal would have a metabolic heat production disproportionate to the surface from which heat could be lost. The total heat production increases with $M_b^{0.75}$, and surface area only with $M_b^{0.67}$. This means that a very large mammal, say an elephant or a rhinoceros, probably *must* be hairless; otherwise, heat loss might not be sufficient for the animal to remain in heat balance in a tropical climate. In a cold climate, however, even an elephant-sized mammal such as a woolly mammoth might need fur.

We can now give a rough summary of the scaling of the thermal problems of the so-called warm-blooded vertebrates, birds and mammals.

Body temperature is body-size-independent, although there are characteristic differences between major groups, such as eutherian and marsupial mammals. Otherwise, body temperature is a scale-independent variable.

In regard to minimizing heat loss, large animals have a double advantage: Their relative surface areas decrease with increasing body size, and in addition, they are better insulated. Conductance decreases with increasing body size more rapidly than specific heat production decreases, resulting in a better tolerance to lower temperatures in the large animal.

Warm-blooded dinosaurs?

What about the giant dinosaurs? There has been a great deal of animated discussion whether or not dinosaurs were "warm-blooded." There are actually two questions involved: One pertains to whether or not dinosaurs maintained a more or less constant body temperature like mammals, and the other concerns whether they had high metabolic rates similar to mammals or lower rates like reptiles in general.

Let us take the second question first. If their metabolic rates were similar to those of modern mammals and fell on an extrapolated mammalian regression line, dinosaurs might have been in serious difficulties or might even have found themselves unable to eliminate the metabolic heat produced. This would have produced cooked dinosaurs, or, to use a more scientific expression, extinction.

Now for the question of body temperature. The sheer bulk of a very large animal will help to smooth out variations in body temperature. When it comes to the size of large dinosaurs, the great bulk would have made it possible to maintain a nearly constant internal temperature for periods of several days, even if they had had low, reptilelike metabolic rates. A model developed by Spotila and associates (1973) predicts that such an animal with a diameter of 1 m would have a thermal time constant of 48 hr, which means that in an equable warm climate this animal could remain essentially "warm-blooded." The thermal inertia would not, however, be sufficient to bridge seasonal variations in ambient temperature.

Dinosaur temperatures cannot be measured directly, but the interpretation of indirect evidence is interesting and highly controversial. A compilation of arguments on both sides can be found in a stimulating symposium volume with the befitting title *A Cold Look at the Warm-Blooded Dinosaurs* (Thomas and Olson, 1980). This book is recommended for those who wish to evaluate for themselves what the evidence is for and against dinosaurs being warm-blooded.

The smallest birds and mammals

We have seen that the smaller the animal, the more difficult it is to balance heat production against the increasing thermal conductance.

The smallest mammals, the shrews, and the smallest birds, the humming-birds, are of about the same size, between 2 and 3 g. Is it a coincidence that the size limit for birds and mammals is similar, or are shrews and hummingbirds up against the same ultimate limit on how small a warm-blooded animal can be? Is it because the heat production of an even smaller animal could not be jacked up high enough to match the heat loss?

We know that this is not true, for many insects can maintain a high body temperature in cold surroundings. Large moths, some beetles, and various bees and bumblebees are, in essence, warm-blooded when they are active. Their flight muscles must be warm in order to produce enough power for flight, and in cold weather they are unable to fly until they have produced enough heat to warm up. The flight muscles are located in the thorax and produce heat by a process similar to shivering in verte-brates until they are warm enough for takeoff. A sphinx moth, for example, requires a muscle temperature of at least 35°C, and if the muscles are at a lower temperature, their speed of contraction is too slow to support flight. Nevertheless, sphinx moths can fly and feed when the air temperature is as low as 10°C, which is made possible through pre-flight heating of the muscles. Retention of heat is aided by the fact that the thorax is covered by long, furry scales that help retain the generated heat during warmup (Heinrich and Bartholomew, 1971).

When these insects are inactive, their body temperature drops toward air temperature, and they become essentially heterothermic. Because of their small body size, they cool off rapidly, which is a further advantage in regard to energy savings. These insects, which weigh from a fraction of a gram to a few grams, have an advantage in their small body size, because both warmup and cooling are rapid; they can be highly oppor-tunistic in the adjustment of their body temperature.

Returning to the hummingbird, we can see an analogous situation. Hummingbirds, when active, have high body temperatures of 38 to 40°C, but at night when they are unable to feed, they permit the body temperature to drop. This is a highly opportunistic strategy, because their metabolic rate is so high that the fuel that they could accumulate during the day might not be sufficient to carry them through the night if they were to maintain a high body temperature. The strategy is surpris-ingly similar to that of the large insects.

It is now evident that insects that are much smaller than humming-birds and shrews are able to keep warm when needed and that their strategy is analogous to that of hummingbirds. We can therefore conclude

Figure 16.1. Specific thermal conductances for large sphingid moths and for birds and mammals. The slope of the regression line for moths is within the confidence limits of the bird-plus-mammal line, and the intercepts at unity body mass are identical for the two lines. Note that the apparently steep slopes of the lines are caused by the different scales on the two axes. From Bartholomew and Epting (1975).

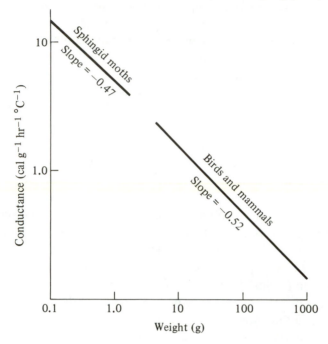

that small body size by itself does not set the lower size limit for warm-blooded animals. Are the conditions for heat loss somehow very different in insects? The answer is negative. The thermal conductance in large moths was compared with that in birds and mammals by Bartholomew and Epting (1975). These investigators calculated the conductance on the basis of the weight of the insect thorax, where the heat-producing organs (flight muscles) are located. The result was that the conductance for large moths fell on a direct extension of the regression line for birds and mammals (Figure 16.1). The slope of the regression line for sphingid moths was −0.47, in the middle of the range reported for mammals and birds (Table 16.2).

These results show, first of all, that a functionally warm-blooded animal can be much smaller than the smallest birds and mammals. Second, such animals operate with a thermal conductance that is scaled to their size exactly as that of birds and mammals. We must conclude that thermal considerations alone are unlikely to determine the ultimate lower limit on body size for birds and mammals.

Is there some other reason that birds and mammals are not smaller? To produce heat, we need both fuel and oxygen. Hummingbirds get most of their energy from nectar, exactly as do the large moths; so fuel supply is an improbable constraint. The oxygen supply systems, on the other hand, are radically different in vertebrates and insects. To transport oxygen from the lungs to the tissues, vertebrates use a convective system consisting of blood pumped by the heart. Insects supply oxygen through air-filled tubes, with oxygen diffusing directly to the tissues where it is consumed. Furthermore, air movement in these tubes is usually improved by active pumping. Even insects as small as fruit flies (*Drosophila*), in spite of their small size, have a highly complex system for pumping air through the tracheal system.

These differences in oxygen supply may well account for the constraints on the minimum size of birds and mammals. The maximum heart rate for shrews and hummingbirds is in the range of 1200 to 1400 beats per minute; that is, one heartbeat lasts between 40 and 50 msec. In that short time, the heart must undergo a full cycle, be filled with venous blood, contract and eject the blood, relax, and be ready for the next filling cycle. It is unlikely that the design of the mammalian heart could permit it to be filled and undergo the full contraction cycle in a shorter time. If increased frequency is impossible, the only way of increasing cardiac output is to increase the heart's size and thus the stroke volume. Indeed, it seems that this solution has already been necessary for both hummingbirds and shrews, which have hearts that are two or three times as large as we would expect from the general scaling of mammalian and bird hearts (see Chapter 11). Apparently, in both shrews and hummingbirds the constraint on increasing heart rate has been compensated for by increasing the size of the heart.

Another possible way to increase the amount of oxygen carried from the heart would be to increase the oxygen capacity of the blood, i.e., increase its hemoglobin content. In this regard, both shrews and hummingbirds have gone as far as feasible: Their bloods have hemoglobin contents as high as are found among any animals. However, there are strict limitations to this solution, because an increased red cell content of the blood (increased hematocrit) causes an inordinate increase in the viscosity of the blood.

The constraints we have mentioned (the contraction time of the heart, the size of the heart, and the viscosity of the blood) all seem to be pushed as far as physiologically possible in both hummingbirds and shrews. It is the different system for oxygen supply in insects that permits these functionally warm-blooded animals to exist at much smaller sizes than the smallest birds and mammals we know about.

Although the food supply is mainly nectar for both hummingbirds and moths, there is a notable difference in the fuel supply to the highly active flight muscles. Whereas the blood of insects does not transport oxygen, it must supply fuel to the heat-producing organs. Perhaps this explains the much higher concentrations of sugars in insect blood than in mammalian and bird blood. Mammalian blood normally has a glucose content of about 1 g/liter; in many insects, especially highly active flying insects, the blood sugar is often 10- or 20-fold higher, thus supplying ample fuel without undue demands on the circulation rate.

In conclusion, we can say with certainty that the minimum size for birds and mammals is not determined by considerations of how small the animal can be and still keep itself warm. The ultimate constraint may be in the design of the oxygen supply system and the limits to the pumping capacity of the heart.

17
Some important concepts

The preceding chapters in this book have dealt with structures and functions and how they are related to body size. We have discussed bones and muscles, energy metabolism and oxygen supply, why time has a different meaning for a mouse and an elephant, and so on. We have considered animals that move about, running and jumping, swimming and flying, and how body size affects the energy cost of locomotion. This is important, for real animals do not sit around doing nothing; they spend much of their time moving about and being active. One important fact is evident: Although comparing animals at rest can provide a great deal of information, it is in the active animal that we are apt to find the limits and constraints on the various functions that make up the whole animal.

Wherever we look at the functions of living organisms, we find that size is important and that a change in size has consequences that require appropriate adjustments or changes. Various functions must be appropriately adjusted; they must be modified as dictated by a change in scale.

Some variables remain size-independent; for example, physical and chemical constants cannot be changed. That is, animals must find the best possible solutions within the existing limitations or rules as determined by the realities of the physical world. When animals meet constraints that set limits to further change in scale, discontinuities in design may solve the problem. These concepts need a brief discussion beyond what was dealt with in the preceding chapters.

Non-scaleable and scale-independent variables

There are many factors that animals can do nothing about. They cannot change physical laws, the field of gravity, the properties of water, the laws of thermodynamics, and such things. These and many other entities can be said to be *non-scaleable,* because they remain unchanged and do not vary with animal size. For example, the force due to gravity on earth is a physical constant that is the same for a mouse and an elephant. Only the science fiction writer can freely manipulate gravity, and modern technology has temporarily removed a few selected astronauts from its immediate influence. For animals on earth, gravity is a fact of reality they must live with.

Non-scaleable entities include all sorts of physical and chemical constants, such as atomic and molecular sizes. Water, the universal solvent for life, and its properties are non-scaleable. This applies to its density, viscosity, heat of vaporization, surface tension, specific heat capacity, and so on. Likewise, physical constants such as diffusion coefficients in air and in water are non-scaleable; that is, these are independent of body size. However, to some extent they can be manipulated; for example, the diffusion of oxygen through tissues can be greatly accelerated by the presence of hemoglobin. We shall return to this question when we discuss how certain constraints can be overcome or circumvented by changes in design, thus forming discontinuities.

Scale-independent variables are different. For example, the properties of many biological materials are similar in large and small animals, such as the physical strength of bones and tendons. These can be said to be scale-independent, a statement based on empirical observation, rather than any fundamental reason that they should be non-scaleable in the sense of physical constants.

When we discussed mammalian muscle, we found that filament diameter, sarcomere length, and filament overlap, and therefore the force that the muscle can exert, all are scale-independent. The muscles of mice and elephants exert the same force per unit cross-sectional area; that is, muscle force is scale-independent. This is again an empirical observation, not a basic property of muscle, because many invertebrate muscles have somewhat different properties that are related to different filament length and structure.

But aren't the muscles of insects many times stronger than our own? An ant can carry in its jaws a prey many times its own weight; it seems to have muscles of inordinate strength. However, measurements show insect muscles to exert the same force per unit cross-sectional area as vertebrate muscles. What makes the ant seem so strong is a simple result of scaling. With decreasing size, the volume or mass of an animal decreases in proportion to the third power of L, but the cross-sectional areas of muscles (which determine the force they can exert) decrease only as the square of L. Thus, the force exerted by muscles, relative to mass, increases in proportion to the decrease in L. This is the reason that the ant appears to have muscles of unmatched strength, although the force that muscle can exert is scale-independent.

Many other structures and functions are scale-independent. We have seen in earlier chapters that lung volume relative to body size of mammals is constant and thus scale-independent. The same applies to blood volume, blood hemoglobin concentration, red cell size, and many other variables. Heart size seems to belong in the same category of scale-independent entities, except that the smallest mammals and birds seem to be pushing up against a constraint on maximal heart rates that requires an increase in relative heart size above the general pattern in their respective groups. Because heartbeat frequency cannot be increased beyond certain limits (apparently about 1300 beats per minute), the only way to increase cardiac output to match the need for oxygen in shrews and hummingbirds is to increase heart size and thus stroke volume.

Optimal design
Many statements in this book are based on the assumption that animal design is optimized. Use of excessive material to build unneeded structures is expensive and useless. Structures must be built to meet maximal demands, but there is no need to go beyond a reasonable safety margin. This leads to optimization without wasteful or unnecessary excess. The same applies to energy: We assume that necessary functions are carried out as economically as possible, and any excess is a waste, not only useless but a disadvantage.

Living animals as they exist are smoothly functioning systems in which both structural material and chemical energy are used with economy. Within the limitations and constraints dictated by physical laws, we can assume that, as far as possible, animals are optimized, because evolution rapidly eliminates the uneconomical and wasteful. Each organism can be considered as representing an optimal design for its size and structure.

It is quite possible, however, that a design that is optimal in one respect is not necessarily optimal in regard to another. For example, a requirement for economy of material may be in conflict with the need for structural strength. This means that different requirements must be balanced, and in order to obtain an optimized solution, we would optimize not each process separately but some combination of the two that would give a unique solution (Rashevsky, 1960). This, in fact, is a process analogous to the optimization concept of modern economic theory.

The next step is to realize that there are not merely two variables to be balanced; the total number of possible interconnections in the living animal is overwhelming. We can just think of the many steps in supplying oxygen to match the metabolic rate. Structures and functions are all interconnected: breathing, lung size and area, diffusion pathways, blood flow, heart, hemoglobin function, capillaries, mitochondria, enzyme concentrations, and so on, in a chain of seemingly unending interdependent variables. Unfortunately, we do not know enough about even the simplest organism to give anything that approaches a complete picture, and we are much too ignorant to include the effects of differing body sizes at more than a greatly simplified level. Nevertheless, the information we have available about size and scaling has already given us a great deal of insight into the possibilities and limitations inherent in problems of body size.

Constraints and discontinuities

A very small organism, say an amoeba, can be supplied with oxygen by diffusion, which suffices for oxygen to reach the entire organism at the required rate. This is not possible for much larger organisms, because diffusion over long distances is a very slow process. This constraint on the oxygen supply can be overcome by using mass transport, convection, to move oxygen. The air in our lungs cannot be renewed by diffusion only, and the lungs are ventilated by convection. Likewise, oxygen cannot be supplied to the various body parts by diffusion, and circulation of blood carries oxygen to the immediate vicinity of the oxygen-consuming cells. These processes are so familiar to us that we may fail to think of them as discontinuities in design, that is, as novel inventions used to overcome the constraints imposed by the slowness of diffusion over long distances.

Another example of a constraint on diffusion that has been overcome by a novel design is the greatly increased diffusion rate for oxygen in hemoglobin-containing tissues. This may be the major function of the hemoglobin (myoglobin) that is found in red muscle.

Constraints on the speed of nerve conduction can likewise be overcome by a novel design. In general, the speed at which an impulse is conducted in a nerve increases with its diameter, as exemplified by the giant axons of squid and some other invertebrates. In these, the high speed of conduction is related to a quick-response mechanism. By a strong contraction of the mantle muscles, a squid can make a quick dart to escape or to capture prey. For the mantle muscle to contract simultaneously, nerve impulses must reach all parts without delay; hence the need for fast conduction and giant axons.

Vertebrate muscles, in contrast, are controlled by nerves that carry hundreds or thousands of single axons. Were these nerves to contain giant axons in order to obtain fast conduction, this would require nerve trunks of inordinate size. In vertebrates, the constraints on the velocity of conduction in nerves have been overcome by a novel design, based on an insulating material (myelin) that permits the very rapid conduction of impulses that is known as saltatory conduction.

These examples show that certain limitations or constraints can be overcome by new inventions or novel designs. This brings us back to the question whether or not the largest and smallest living organisms we know about are actual limits on what is possible. This was discussed in Chapter 1, where organisms of widely different sizes were listed in Table 1.1. There are cogent reasons to believe that the smallest and largest organisms in the list represent approximate limits to the possible size of animals under conditions that prevail on our planet. A question that remains unanswered, however, concerns the limit to the size of mammals. We were unable to answer the question of what limits the size of the largest land-living mammal. Likewise, we have been unable to determine with certainty if shrews and hummingbirds that weigh 2 or 3 g represent the smallest possible warm-blooded vertebrates. It appears that a different design of the oxygen supply system, as in insects, permits much smaller animals such as moths and bumblebees to operate with a high and well-regulated body temperature. That is, these animals are in principle what we call warm-blooded, but we do not know for certain if there are constraints that prevent the existence of equally small warm-blooded birds and mammals.

Ecological implications

What can we learn about animal size and its biological implications, beyond the functions we have discussed? Thus far we have talked about whole animals and their parts, and nothing has been said about the environment in which they live. Major parts of this book have focused on energy metabolism and metabolic rates and related structures and functions such as lungs, blood, heart, and so forth. Perhaps this is because energy metabolism has been better studied than most other physiological functions, and we also tend to regard metabolic rate as something very fundamental that determines a lot of other processes.

We realize that metabolic rate establishes the rates of food intake, excretion, and a host of other functions. To remain in a steady state, the turnover of energy and nutrients should be matched by the ingestion rate of foods. In ecological terminology this can simply be equated with *energy flux,* which is a central theme in ecology.

The need for energy, or the necessary ingestion rate, is a determining factor for *animal abundance and density*. It is well known in ecology that animal density is inversely related to body size, among both herbivores and predators. This has profound implications for animal distribution and the allometric description of home-range areas. Food extraction from the environment is inextricably related to the use of resources and the optimization of population densities, relationships that are at the very foundation of modern ecology.

Within the discipline of scaling, perhaps more important than any other function, except metabolic rate, is the energy used for animal locomotion. The clear and regular size dependence in the use of energy sources for moving about has ecological consequences that are not easily sorted out. We saw in Chapter 14 that it is cheaper to move 1 kg of a large animal than the same mass of a small animal over a given distance. Evidently, it is cheaper to be big, and bigger is better! But a horse must move over greater distances than a mouse to find enough food. And if resources are limited (nearly always), it is to the advantage of an individual mouse that it needs less food than a horse. A hectare of grassland can support a huge population of field mice, but no more than one or two horses. Evidently, smaller is better! The conclusion is that answers are not simple, and animal size is a complex subject that is of fundamental importance to general ecological principles.

This is not the place for a further discussion of the ecological implications of animal size. The subject has been reviewed in a recent monograph that discusses subjects such as growth and reproduction, mass and energy flow, animal abundance, and other ecological relations from the viewpoint of animal size and scaling (Peters, 1983).

Every biologist who is interested in the problems of body size should be familiar with a few major works. Although others may be equally important, and many have already been quoted in earlier chapters, I shall mention the following. First is the charmingly written book *On Growth and Form,* by D'Arcy Thompson (1961, first published in 1917), which is most readable and enjoyable. Kleiber's book *The Fire of Life* (1961) is an often-quoted classic. A review of broad biological interest is "Allometry and Size in Ontogeny and Phylogeny," by Stephen Gould (1966). Finally, I wish to mention a highly successful symposium, *Scale Effects in Animal Locomotion,* edited by Pedley (1977). These authors have each covered different and important aspects of a fascinating field of biology, the importance of size and scale.

Appendixes

A

Symbols used

a = proportionality coefficient
b = exponent, allometric exponent, slope
A = area, surface area
F = force
f = frequency
H = heat
l, L = length
M = mass
M_b = body mass
P = power, metabolic rate
RME = residual mass exponent
S = surface area
T = temperature
t = time
U = speed, velocity
V = volume
$\dot{V}_{O_2}$ = rate of oxygen consumption
w = weight
W = work
$*$ = specific, divided by mass; for example, specific power, power per unit mass = P^*
M L T = dimensions of mass, length, time

Other symbols are defined where they are used in the text.

B

The allometric equation

The allometric equation has the general form $y = ax^b$, and in logarithmic form: $\log y = \log a + b \log x$.

In problems of scaling, in which structures or functions are related to animal size, it is customary to consider body size the independent variable (x).

Let M_b represent body mass in the allometric equation

$$y = a M_b^b$$

In this equation, the exponent, b, is called the *body-mass exponent*. This is the same as the *slope* of the straight line that represents the allometric equation in a plot on logarithmic coordinates.

The *proportionality coefficient, a*, is identical with the intercept of the regression line at unity body mass, or $M_b = 1$. This is because the number 1 raised to any power remains 1, thus giving $y = a$.

Computer plots are often obtained when data are processed for computation of statistical parameters and regression lines. Many such computer plots have a serious shortcoming or disadvantage in that the two axes have the dividing marks at different distances apart. When the coordinates have different scales, the slope of the regression line is distorted and therefore is not perceived correctly at first glance. Thus, if the computer plot has expanded the y axis disproportionately relative to the x axis, the regression line appears much steeper than in reality it should be. If the computer expands or contracts the two axes to different degrees, this results in a misleading appearance of the slopes of regression lines. Figure 16.1 is an example. For any plot of a regression, it is recommended that the divisions on the two axes be equal distances apart, so that the true slopes of regression lines will be immediately apparent.

The body mass, or body weight, is customarily plotted on the abscissa (x axis). This is not simply because body mass can be determined with greater accuracy than many other variables, but primarily because we are interested in how a function, say metabolic rate, changes with body size. It would be awkward to consider that body size changes as a function of metabolic rate. Thus, in scaling, body size should always be plotted on the abscissa.

Some caution is necessary in recording body mass or body weight. For example, is the content of the digestive tract (up to 20 or 25% of body weight for a ruminant) part of the body weight? Are large fat deposits that have a very low metabolic rate to be considered an appropriate and integral part of the metabolizing body mass? What body mass should be used for a clam that has thick, heavy shells, weighing perhaps several times as much as the living part of the animal? What body mass should we use for a jellyfish, which contains mostly water and very little active metabolizing tissue?

There are no simple answers to these questions. An attempt at finding a uniform basis for comparisons consists in using the total protein content of an organism. This has some merit, but if total nitrogen content is used to estimate protein mass, we are in trouble with animals such as sharks, which retain large amounts of nitrogen-containing solutes for osmoregulatory purposes. The answer is, at this time, that comparisons should be made between groups of similar animals, that some common sense should be used, and that we do not have all the answers.

C

Recalculation of equations according to units used for body mass

The body mass of animals may be expressed in grams or in kilograms, or even in other units. A change from one set of units to another changes the numerical value of the proportionality coefficient (a).

If a given equation expresses M_b in grams, a recalculation to M_b in kilograms consists in the following. The right-hand side of the equation is multiplied by the expression $1000^b/1000^b$ (which equals 1, and thus leaves the equation unchanged, but permits the following change).

When changing from grams to kilograms,

$$y = ax^b \cdot \frac{1000^b}{1000^b}$$

$$= (a \cdot 1000^b) \cdot \frac{x^b}{1000^b}$$

$$= a \cdot 1000^b \cdot \left(\frac{x}{1000}\right)^b$$

That is, the numerical value of x is divided by 1000, and the numerical value of a is multiplied by 1000^b.

When changing from kilograms to grams, the corresponding operation yields

$$y = \frac{a}{1000^b} \cdot (1000\,x)^b$$

D

Algebraic rules for
operating with expressions
that contain powers and roots

$$x^a \cdot x^b = x^{(a+b)}$$

$$\frac{x^a}{x^b} = x^{(a-b)}$$

$$(x^a)^b = x^{ab}$$

$$x^{-a} = \frac{1}{x^a}$$

$$x^{1/a} = \sqrt[a]{x}$$

$$x^{a/b} = \sqrt[b]{x^a}$$

$$x^0 = 1$$

$$\text{if } y = x^a, \quad \text{then } x = \sqrt[a]{y}$$

E

Dimensional formulas for some commonly used physical quantities in the M L T system

M is the dimensional symbol for mass, L for length, and T for time. Dimensional symbols can be multiplied or divided, but cannot meaningfully be added or subtracted.

Velocity, speed	$L T^{-1}$
Acceleration	$L T^{-2}$
Frequency	T^{-1}
Density	$M L^{-3}$
Momentum	$M L T^{-1}$
Force	$M L T^{-2}$
Energy, work	$M L^2 T^{-2}$
Power	$M L^2 T^{-3}$
Pressure, stress	$M L^{-1} T^{-2}$
Viscosity (dynamic)	$M L^{-1} T^{-1}$

References

Alexander, R. McN. (1968) *Animal Mechanics.* Seattle: University of Washington Press. 346 pp.

Alexander, R. McN., Jayes, A. S., Maloiy, G. M. O., and Wathuta, E. M. (1979*a*) Allometry of the limb bones of mammals from shrews (*Sorex*) to elephant (*Loxodonta*). *J. Zool., Lond. 189*:305–14.

– (1981) Allometry of the leg muscles of mammals. *J. Zool., Lond. 194*:539–52.

Alexander, R. McN., Maloiy, G. M. O., Hunter, B., Jayes, A. S., and Nturibi, J. (1979*b*) Mechanical stresses in fast locomotion of buffalo (*Syncerus caffer*) and elephant (*Loxodonta africana*). *J. Zool., Lond. 189*:135–44.

Altman, P. L., and Dittmer, D. S. (1961) *Blood and Other Body Fluids.* Washington, D.C.: Federation of American Societies for Experimental Biology. 539 pp.

– (1964) *Biology Data Book.* Washington, D.C.: Federation of American Societies for Experimental Biology. 633 pp.

– (1968) *Metabolism.* Bethesda: Federation of American Societies for Experimental Biology. 737 pp.

– (1971) *Respiration and Circulation.* Bethesda: Federation of American Societies for Experimental Biology. 930 pp.

Andersen, H. T. (ed.) (1969) *The Biology of Marine Mammals.* New York: Academic Press. 511 pp.

Anderson, J. F. (1970) Metabolic rates of spiders. *Comp. Biochem. Physiol. 33*:51–72.

– (1974) Responses to starvation in the spiders *Lycosa lenta* Hentz and *Filistata hibernalis* (Hentz). *Ecology 55*:576–85.

Anderson, J. F., Rahn, H., and Prange, H. D. (1979) Scaling of supportive tissue mass. *Quart. Rev. Biol. 54*:139–48.

Ar, A., Paganelli, C. V., Reeves, R. B., Greene, D. G., and Rahn, H. (1974) The avian egg: water vapor conductance, shell thickness, and functional pore area. *Condor 76*:153–8.

Ar, A., Rahn, H., and Paganelli, C. V. (1979) The avian egg: mass and strength. *Condor 81*:331–7.

Aschoff, J. (1981) Thermal conductance in mammals and birds: its dependence on body size and circadian phase. *Comp. Biochem. Physiol. 69A*:611–19.

Aschoff, J., Günther, B., and Kramer, K. (1971) *Energiehaushalt und Temperatur-regulation.* Munich: Urban & Schwarzenberg. 196 pp.

Aschoff, J., and Pohl, H. (1970*a*) Der Ruheumsatz von Vögeln als Funktion der Tageszeit und der Körpergrösse. *J. Ornithol. 3*:38–48.

– (1970*b*) Rhythmic variations in energy metabolism. *Fed. Proc. 29*:1541–52.

Bainbridge, R. (1958) The speed of swimming of fish as related to size and to the frequency and amplitude of the tail beat. *J. Exp. Biol. 35*:109–33.

Bakker, R. T. (1972) Locomotor energetics of lizards and mammals compared. *Physiologist 15*:76.

Ballard, F. J., Hanson, R. W., and Kronfeld, D. S. (1969) Gluconeogenesis and lipogenesis in tissue from ruminant and nonruminant animals. *Fed. Proc. 28*:218–31.

Banchero, N., Grover, R. F., and Will, J. A. (1971) Oxygen transport in the llama (*Lama glama*). *Resp. Physiol. 13*:102–15.

Barlow, G. W. (1961) Intra- and interspecific differences in rate of oxygen consumption in gobiid fishes of the genus *Gillichthys*. *Biol. Bull. 121*:209–29.

Bartels, H. (1964) Comparative physiology of oxygen transport in mammals. *Lancet 1964*:599–604.

– (1980) Aspekte des Gastransports bei Säugetieren mit hoher Stoffwechselrate. *Verh. Dtsch. Zool. Ges. 1980*:188–201.

– (1982) Metabolic rate of mammals equals the 0.75 power of their body weight. *Exp. Biol. Med. 7*:1–11.

Bartholomew, G. A., and Epting, R. J. (1975) Allometry of post-flight cooling rates in moths: a comparison with vertebrate homeotherms. *J. Exp. Biol. 63*:603–13.

Bartholomew, G. A., and Tucker, V. A. (1964) Size, body temperature, thermal conductance, oxygen consumption, and heart rate in Australian varanid lizards. *Physiol. Zool. 37*:341–54.

Baudinette, R. V. (1978). Scaling of heart rate during locomotion in mammals. *J. Comp. Physiol. 127*:337–42.

Beamish, F. W. H. (1964) Respiration of fishes with special emphasis on standard oxygen consumption. II. Influence of weight and temperature on respiration of several species. *Can. J. Zool. 42*:177–88.

– (1978) Swimming capacity. In *Fish Physiology, Vol. 7* (W. S. Hoar and D. J. Randall, eds.), pp. 101–87. New York: Academic Press.

Beamish, F. W. H., and Mookherjii, P. S. (1964) Respiration of fishes with special emphasis on standard oxygen consumption. I. Influence of weight and temperature on respiration of goldfish, *Carassius auratus* L. *Can. J. Zool. 42*:161–75.

Benedict, F. G. (1934) Die Oberflächenbestimmung verschiedener Tiergattungen. *Ergeb. Physiol. 36*:300–46.

– (1938) *Vital Energetics: A Study in Comparative Basal Metabolism.* Washington, D.C.: Carnegie Institute of Washington. 215 pp.

Bennet-Clark, H., and Alder, G. M. (1979) The effect of air resistance on the jumping performance of insects. *J. Exp. Biol. 82*:105–21.

Bennett, A. F., and Dawson, W. R. (1976) Metabolism. In *Biology of the Reptilia, Physiology A, Vol. 5* (C. Gans and W. R. Dawson, eds.), pp. 127–223. New York: Academic Press.

Berger, M., and Hart, J. S. (1974) Physiology and energetics of flight. In *Avian Biology, Vol. 4* (D. S. Farner and J. R. King, eds.), pp. 415–77. New York: Academic Press.

Bergmann, C. (1847) Ueber die Verhältnisse der Wärmeökonomie der Thiere zu ihrer Grösse. *Göttinger Studien*, pp. 595–708.

Bertalanffy, L. von, and Pirozynski, W. J. (1951a) Tissue respiration and body size. *Science 113*:599–600.

– (1951b) Tissue respiration and body size. *Science 114*:306–7.

Biewener, A. A. (1982) Bone strength in small mammals and bipedal birds: Do safety factors change with body size? *J. Exp. Biol. 98*:289–301.

– (1983) Locomotory stresses in the limb bones of two small mammals: the ground squirrel and chipmunk. *J. Exp. Biol. 103*:131–54.

Bland, D. K., and Holland, R. A. B. (1977) Oxygen affinity and 2,3-diphosphoglycerate in blood of Australian marsupials of differing body size. *Resp. Physiol. 31*:279–90.

Bradley, S. R., and Deavers, D. R. (1980) A re-examination of the relationship between thermal conductance and body weight in mammals. *Comp. Biochem. Physiol.* *65A*:465–76.

Brett, J. R. (1964) The respiratory metabolism and swimming performance of young sockeye salmon. *J. Fish. Res. Bd. Canada 21*:1183–226.

– (1965) The relation of size to rate of oxygen consumption and sustained swimming speed of sockeye salmon (*Oncorhynchus nerka*). *J. Fish. Res. Bd. Canada 22*:1491–7.

Bridgman, P. W. (1937) *Dimensional Analysis*. New Haven: Yale University Press. 113 pp.

Brody, S. (1945) *Bioenergetics and Growth, with Special Reference to the Efficiency Complex in Domestic Animals*. New York: Reinhold. 1023 pp. (reprinted 1964. Darien, CT: Hafner).

Brody, S., and Elting, E. C. (1926) Growth and development with special reference to domestic animals. II. A new method for measuring surface area and its utilization to determine the relation between growth in surface area and growth in weight and skeletal growth in dairy cattle. *Univ. Missouri Agric. Exp. Sta. Res. Bull. 89*:1–18.

Brody, S., Procter, R. C., and Ashworth, U. S. (1934) Basal metabolism, endogenous nitrogen, creatinine and neutral sulphur excretions as functions of body weight. *Univ. Missouri Agric. Exp. Sta. Res. Bull. 220*:1–40.

Burke, J. D. (1966) Vertebrate blood oxygen capacity and body weight. *Nature 212*:46–8.

Burstein, A. H., Currey, J. D., Frankel, V. H., and Reilly, D. T. (1972) The ultimate properties of bone tissue: the effects of yielding. *J. Biomechanics 5*:35–44.

Butler, P. J., West, N. H., and Jones, D. R. (1977) Respiratory and cardiovascular responses of the pigeon to sustained, level flight in a wind-tunnel. *J. Exp. Biol. 71*:7–26.

Butler, P. J., and Woakes, A. J. (1980) Heart rate, respiratory frequency and wing beat frequency of free flying barnacle geese *Branta leucopsis. J. Exp. Biol. 85*:213–26.

Calder, W. A. (1968) Respiratory and heart rates of birds at rest. *Condor 70*:358–65.

– (1976) Aging in vertebrates: allometric considerations of spleen size and lifespan. *Fed. Proc. 35*:96–7.

Calder, W. A., and King, J. R. (1974) Thermal and caloric relations of birds. In *Avian Biology, Vol. 4* (D. S. Farner and J. R. King, eds.), pp. 259–413. New York: Academic Press.

Calder, W. A., and Schmidt-Nielsen, K. (1968) Panting and blood carbon dioxide in birds. *Amer. J. Physiol. 215*:477–82.

Carey, F. G., and Teal, J. M. (1966) Heat conservation in tuna fish muscle. *Proc. Natl. Acad. Sci. U.S.A. 56*:1464–9.

– (1969) Regulation of body temperature by the bluefin tuna. *Comp. Biochem. Physiol. 28*:205–13.

Carrel, J. E., and Heathcote, R. D. (1976) Heart rate in spiders: influence of body size and foraging energetics. *Science 193*:148–50.

Casey, T. M. (1981) A comparison of mechanical and energetic estimates of flight cost for hovering sphinx moths. *J. Exp. Biol. 91*:117–29.

Clark, A. J. (1927) *Comparative Physiology of the Heart*. New York. Macmillan. 157 pp.

Cohen, Y., Robbins, C. T., and Davitt, B. B. (1978) Oxygen utilization by elk calves during horizontal and vertical locomotion compared to other species. *Comp. Biochem. Physiol. 61A*:43–8.

Colbert, E. H. (1962) The weights of dinosaurs. *Amer. Mus. Novitates 2076*:1–16.

Crile, G., and Quiring, D. P. (1940) A record of the body weight and certain organ and gland weights of 3,690 animals. *Ohio J. Sci. 40*:219–59.

Currey, J. D. (1967) The failure of exoskeletons and endoskeletons. *J. Morphol. 123*:1–16

Czopek, J. (1965) Quantitative studies on the morphology of respiratory surfaces in amphibians. *Acta Anat. 62*:296–323.

Dawson, T. J., and Hulbert, A. J. (1970) Standard metabolism, body temperature, and surface areas of Australian marsupials. *Amer. J. Physiol. 218*:1233–8.

Dedrick, R. L., Bischoff, K. B., and Zaharko, D. S. (1970) Interspecies correlation of plasma concentration history of methotrexate (NSC-740). *Cancer Chemother. Rep. 54*:95–101.

Dhindsa, D. S., Hoversland, A. S., and Metcalfe, J. (1971) Respiratory functions of armadillo blood. *Resp. Physiol. 13*:198–208.

Dmi'el, R. (1972) Relation of metabolism to body weight in snakes. *Copeia 1972*:179–81.

Dosse, G. (1937) Vergleichende Gewichtsuntersuchungen am Vogelskelett. *Zool. Jahr. Anat. 63*:299–350.

Drabkin, D. L. (1950) The distribution of the chromoproteins, hemoglobin, myoglobin, and cytochrome c, in the tissues of different species, and the relationship of the total content of each chromoprotein to body mass. *J. Biol. Chem. 182*:317–33.

Drorbaugh, J. E. (1960) Pulmonary function in different animals. *J. Appl. Physiol. 15*:1069–72.

Dunaway, P. B., and Lewis, L. L. (1965) Taxonomic relation of erythrocyte count, mean corpuscular volume, and body-weight in mammals. *Nature 205*:481–4.

Economos, A. C. (1979) Gravity, metabolic rate and body size of mammals. *Physiologist 22(Suppl.)*:S-71.

Edwards, N. A. (1975) Scaling of renal functions in mammals. *Comp. Biochem. Physiol. 52A*:63–6.

Edwards, R. R. C., Finlayson, D. M., and Steele, J. H. (1969) The ecology of O-group plaice and common dabs in Loch Ewe. II. Experimental studies of metabolism. *J. Exp. Mar. Biol. Ecol. 3*:1–17.

Emmett, B., and Hochachka, P. W. (1981) Scaling of oxidative and glycolytic enzymes in mammals. *Resp. Physiol. 45*:261–72.

Fedak, M. A., Pinshow, B., and Schmidt-Nielsen, K. (1974) Energy cost of bipedal running. *Amer. J. Physiol. 227*:1038–44.

Fedak, M. A., and Seeherman, H. J. (1979) Reappraisal of energetics of locomotion shows identical cost in bipeds and quadrupeds including ostrich and horse. *Nature 282*:713–16.

Feduccia, A. (1980) *The Age of Birds*. Cambridge: Harvard University Press. 196 pp.

Feldman, H. A., and McMahon, T. A. (1983) The 3/4 mass exponent for energy metabolism is not a statistical artifact. *Resp. Physiol. 52*:149–63.

Field, J., II, Belding, H. S., and Martin, A. W. (1939) An analysis of the relation between basal metabolism and summated tissue respiration in the rat. I. The post-pubertal albino rat. *J. Cell. Comp. Physiol. 14*:143–57.

Folkow, B., and Neil, E. (1971) *Circulation*. New York: Oxford University Press. 593 pp.

Fry, F. E. J. (1957) The aquatic respiration of fish. In *The Physiology of Fishes, Vol. 1* (M. E. Brown, ed.), pp. 1–63. New York: Academic Press.

Galilei, G. (1637) *Dialogues Concerning Two New Sciences* (translated by H. Crew and A. De Salvio). New York: Macmillan, 1914. 300 pp.

Galvão, P. E., Tarasantchi, J., and Guertzenstein, P. (1965) Heat production of tropical snakes in relation to body weight and body surface. *Amer. J. Physiol. 209*:501–6.

Gehr, P., Mwangi, D. K., Ammann, A., Maloiy, G. M. O., Taylor, C. R., and Weibel, E. R. (1981) Design of the mammalian respiratory system. V. Scaling morphometric pulmonary diffusing capacity to body mass: wild and domestic mammals. *Resp. Physiol. 44*:61–86.

Gould, S. J. (1966) Allometry and size in ontogeny and phylogeny. *Biol. Rev. 41*:587–640.

Grafe, E. (1925) Probleme der Gewebsatmung. *Deutsch Med. Wochenschr. 51*:640–2.

Grafe, E., Reinwein, H., and Singer (1925) Studien über Gewebsatmung. II. Mitteilung: Die Atmung der überlebenden Warmblüterorgane. *Biochem. Zeitschr. 165*:102–17.

Graham, J. B. (1973) Heat exchange in the black skipjack, and the blood-gas relationship of warm-bodied fishes. *Proc. Natl. Acad. Sci. U.S.A. 70*:1964–7.

Graham, J. M. (1949) Some effects of temperature and oxygen pressure on the metabolism and activity of the speckled trout, *Salvelinus fontinalis. Can. J. Res. 27(Section D)*:270–88.

Grande, F., and Taylor, H. L. (1965) Adaptive changes in the heart, vessels, and patterns of control under chronically high loads. In *Handbook of Physiology, Sec. 2: Circulation, Vol. 3* (W. F. Hamilton and P. Dow, eds.), pp. 2615–77. Washington, D.C.: American Physiological Society.

Granger, W., and Gregory, W. K. (1935) A revised restoration of the skeleton of *Baluchitherium,* gigantic fossil rhinoceros of central Asia. *Amer. Mus. Novitates 787*:1–3.

Gray, I. E. (1954) Comparative study of the gill area of marine fishes. *Biol. Bull. 107*:219–25.

Greenewalt, C. H. (1962) Dimensional relationships for flying animals. *Smithsonian Misc. Collections 144*:1–46.

– (1975*a*) Could pterosaurs fly? *Science 188*:676.

– (1975*b*) The flight of birds. The significant dimensions, their departure from the requirements of dimensional similarity, and the effect on flight aerodynamics of that departure. *Trans. Amer. Phil. Soc. 65(4)*:1–67.

Gregory, W. K. (1951) *Evolution Emerging. A Survey of Changing Patterns from Primeval Life to Man, Vols. 1 and 2.* New York: Macmillan. 736 and 1013 pp.

Grubb, B. (1983) Allometric relations of cardiovascular function in birds. *Amer. J. Physiol. 245*:H567–72.

Hall, F. G. (1966) Minimal utilizable oxygen and the oxygen dissociation curve of blood of rodents. *J. Appl. Physiol. 21*:375–8.

Hall, F. H., Dill, D. B., and Barron, E. S. G. (1936) Comparative physiology in high altitudes. *J. Cell. Comp. Physiol. 8*:301–13.

Hall-Craggs, E. C. B. (1965) An analysis of the jump of the lesser galago (*Galago senegalensis*). *J. Zool. 147*:20–9.

Hart, J. S. (1956) Seasonal changes in insulation of the fur. *Can. J. Zool. 34*:53–7.

Hart, J. S., and Roy, O. Z. (1966) Respiratory and cardiac responses to flight in pigeons. *Physiol. Zool. 39*:291–305.

Hartman, F. A. (1955) Heart weight in birds. *Condor 57*:221–38.

Heglund, N. C., Taylor, C. R., and McMahon, T. A. (1974) Scaling stride frequency and gait to animal size: mice to horses. *Science 186*:1112–3.

Heinrich, B., and Bartholomew, G. A. (1971) An analysis of pre-flight warm-up in the sphinx moth, *Manduca sexta. J. Exp. Biol. 55*:223–39.

Hemmingsen, A. M. (1950) The relation of standard (basal) energy metabolism to total fresh weight of living organisms. *Rep. Steno Mem. Hosp. (Copenhagen) 4*:1–58.

– (1960) Energy metabolism as related to body size and respiratory surfaces, and its evolution. *Rep. Steno Mem. Hosp. (Copenhagen) 9*:1–110.

Herreid, C. F., II, and Kessel, B. (1967) Thermal conductance in birds and mammals. *Comp. Biochem. Physiol. 21*:405–14.

Heusner, A. A. (1982) Energy metabolism and body size. I. Is the 0.75 mass exponent of Kleiber's equation a statistical artifact? *Resp. Physiol. 48*:1–12.

– (1983) Body size, energy metabolism, and the lungs. *J. Appl. Physiol. (Resp. Environ. Exercise Physiol.) 54*:867–73.

Hill, A. V. (1950) The dimensions of animals and their muscular dynamics. *Proc. Roy. Inst. G.B. 34*:450–71.

Hills, B. A., and Hughes, G. M. (1970) A dimensional analysis of oxygen transfer in the fish gill. *Resp. Physiol. 9*:126–40.

Hinds, D. S., and Calder, W. A. (1971) Tracheal dead space in the respiration of birds. *Evolution 25*:429–40.

Holt, J. P., Rhode, E. A., and Kines, H. (1968) Ventricular volumes and body weight in mammals. *Amer. J. Physiol. 215*:704–15.

Hughes, G. M. (1966) The dimensions of fish gills in relation to their function. *J. Exp. Biol. 45*:177–95.

Hughes, G. M., and Gray, I. E. (1972) Dimensions and ultrastructure of toadfish gills. *Biol. Bull. 143*:150–61.

Hughes, G. M., and Morgan, M. (1973) The structure of fish gills in relation to their respiratory function. *Biol. Rev. 48*:419–75.

Hutchison, V. H., Whitford, W. G., and Kohl, M. (1968) Relation of body size and surface area to gas exchange in anurans. *Physiol. Zool. 41*:65–85.

Huxley, J. S. (1927) On the relation between egg-weight and body-weight in birds. *J. Linnean Soc., Zoology 36*:457–66.

Irving, L., and Krog, J. (1954) Body temperatures of arctic and subarctic birds and mammals. *J. Appl. Physiol. 6*:667–80.

Iversen, J. A. (1972) Basal energy metabolism of mustelids. *J. Comp. Physiol. 81*:341–4.

Jansky, L. (1959) Working oxygen consumption in two species of wild rodents (*Microtus arvalis, Clethrionomys glareolus*). *Physiol. Bohemoslov. 8*:472–8.

– (1961) Total cytochrome oxidase activity and its relation to basal and maximal metabolism. *Nature 189*:921–2.

– (1963) Body organ cytochrome oxidase activity in cold- and warm-acclimated rats. *Can. J. Biochem. Physiol. 41*:1847–54.

Jensen, T. F., and Holm-Jensen, I. (1980) Energetic cost of running in workers of three ant species, *Formica fusca* L., *Formica rufa* L., and *Camponotus herculeanus* L. (Hymenoptera, Formicidae). *J. Comp. Physiol. 137*:151–6.

Jerison, H. J. (1969) Brain evolution and dinosaur brains. *Amer. Natur. 103*:575–88.

– (1970) Gross brain indices and the analysis of fossil endocasts. In *Advances in Primatology, Vol. 1: The Primate Brain* (C. R. Noback and W. Montagna, eds.), pp. 225–44. New York: Appleton-Century-Crofts.

Job, S. V. (1955) The oxygen consumption of *Salvelinus fontinalis*. *Publ. Ontario Fish. Res. Lab. 73(Univ. Toronto Studies, Biol. Ser., No. 61)*:1–39.

Jürgens, K. D., Bartels, H., and Bartels, R. (1981) Blood oxygen transport and organ weights of small bats and small nonflying mammals. *Resp. Physiol. 45*:243–60.

Kinnear, J. E., and Brown, G. D. (1967) Minimum heart rates of marsupials. *Nature 215*:1501.

Kleiber, M. (1932) Body size and metabolism. *Hilgardia 6*:315–53.

– (1941) Body size and metabolism of liver slices *in vitro*. *Proc. Soc. Exp. Biol. Med. 48*:419–23.

– (1961) *The Fire of Life. An Introduction to Animal Energetics*. New York: Wiley. 454 pp.

Krebs, H. A. (1950) Body size and tissue respiration. *Biochim. Biophys. Acta 4*:249–69.

Krogh, A. (1929) *The Anatomy and Physiology of Capillaries* (ed. 2). New Haven: Yale University Press.

Kunkel, H. O., Spalding, J. F., de Franciscis, G., and Futrell, M. F. (1956) Cytochrome oxidase activity and body weight in rats and in three species of large animals. *Amer. J. Physiol. 186*:203–6.

Lahiri, S. (1975) Blood oxygen affinity and alveolar ventilation in relation to body weight in mammals. *Amer. J. Physiol. 229*:529–36.

Langman, V. A., Baudinette, R. V., and Taylor, C. R. (1981) Maximum aerobic capacity of wild and domestic canids compared. *Fed. Proc. 40*:432 (abstract 1142).

Larimer, J. L. (1959) Hemoglobin concentration and oxygen capacity of mammalian blood. *J. Elisha Mitchell Sci. Soc. 75*:174–7.

Lasiewski, R. C. (1964) Body temperatures, heart and breathing rate, and evaporative water loss in hummingbirds. *Physiol. Zool. 37*:212–23.

Lasiewski, R. C., and Calder, W. A., Jr. (1971) A preliminary allometric analysis of respiratory variables in resting birds. *Resp. Physiol. 11*:152–66.

Lasiewski, R. C., and Dawson, W. R. (1967) A re-examination of the relation between standard metabolic rate and body weight in birds. *Condor 69*:13–23.

– (1969) Calculation and miscalculation of the equations relating avian standard metabolism to body weight. *Condor 71*:335–6.

Lawson, D. A. (1975) Pterosaur from the latest Cretaceous of West Texas: discovery of the largest flying creature. *Science 187*:947–8.

Lenfant, C., and Sullivan, K. (1971) Adaption to high altitude. *New Engl. J. Med. 284*:1298–309.

Lighthill, J. (1974) Aerodynamic aspects of animal flight. British Hydromechanics Research Association, 5th Fluid Science Lecture, June 1974. 30 pp.

Lindstedt, S. L., and Calder, W. A. (1976) Body size and longevity in birds. *Condor 78*:91–4.

– (1981) Body size, physiological time, and longevity of homeothermic animals. *Quart. Rev. Biol. 56*:1–16.

Lutz, P. L., Longmuir, I. S., and Schmidt-Nielsen, K. (1974) Oxygen affinity of bird blood. *Resp. Physiol. 20*:325–30.

MacMillen, R. E., and Nelson, J. E. (1969) Bioenergetics and body size in dasyurid marsupials. *Amer. J. Physiol. 217*:1246–51.

Maina, J. N., and Settle, J. G. (1982) Allometric comparisons of some morphometric parameters of avian and mammalian lungs. *J. Physiol. (Lond.) 330*:28P.

Mallouk, R. S. (1975) Longevity in vertebrates is proportional to relative brain weight. *Fed. Proc. 34*:2102–3.

– (1976) Author's reply (to: Aging in vertebrates: allometric considerations of spleen size and lifespan, by W. A. Calder III). *Fed. Proc. 35*:97–8.

Margaria, R. (1976) *Biochemics and Energetics of Muscular Exercise.* Oxford: Clarendon Press. 146 pp.

Martin, A. W., and Fuhrman, F. A. (1955) The relationship between summated tissue respiration and metabolic rate in the mouse and dog. *Physiol. Zool. 28*:18–34.

Martin, C. J. (1903) Thermal adjustment and respiratory exchange in monotremes and marsupials. A study in the development of homeothermism. *Phil. Trans. Roy. Soc. London B 195*:1–37.

Mathieu, O., Krauer, R., Hoppeler, H., Gehr, P., Lindstedt, S. L., Alexander, R. McN., Taylor, C. R., and Weibel, E. R. (1981) Design of the mammalian respiratory system. VII. Scaling mitochondrial volume in skeletal muscle to body mass. *Resp. Physiol. 44*:113–28.

McMahon, T. (1973) Size and shape in biology. *Science 179*:1201–4.

– (1975a) The mechanical design of trees. *Sci. Amer. 233(1)*:92–102.

– (1975b) Allometry and biomechanics: limb bones in adult ungulates. *Amer. Nat. 109*:547–63.

McMahon, T., and Kronauer, R. E. (1976) Tree structures: deducing the principle of mechanical design. *J. Theor. Biol. 59*:443–66.

Meeh, K. (1879) Oberflächenmessungen des menschlichen Körpers. *Zeit. Biol. 15*:425–58.

Morowitz, H. J. (1966) The minimum size of cells. In *Principles of Biomolecular Organization* (G. E. W. Wolstenholme and M. O'Connor, eds.), pp. 446–59. London: J. and A. Churchill.

Morrison, P. R., and Ryser, F. A. (1952) Weight and body temperature in mammals. *Science 116*:231–2.

Morrison, P. R., Ryser, F. A., and Dawe, A. R. (1959) Studies on the physiology of the masked shrew *Sorex cinereus. Physiol. Zool. 32*:256–71.

Muir, B. S., and Hughes, G. M. (1969) Gill dimensions for three species of tunny. *J. Exp. Biol. 51*:271–85.

Munro, H. N. (1969) Evolution of protein metabolism in mammals. In *Mammalian Protein Metabolism, Vol. 3* (H. N. Munro, ed.), pp. 133–82. New York: Academic Press.

O'Neil, J. J., and Leith, D. E. (1980) Lung diffusing capacity scaled in mammals from 25 g to 500 kg. *Fed. Proc. 39*:972.

Ostrom, J. H., and McIntosh, J. S. (1966) *Marsh's Dinosaurs. The Collections from Como Bluff.* New Haven: Yale University Press. 388 pp.

Pace, N., and Smith, A. H. (1981) Gravity, and metabolic scale effects in mammals. *Physiologist 24(Suppl.)*:S37–40.

Paganelli, C. V., Olszowka, A., and Ar, A. (1974) The avian egg: surface area, volume, and density. *Condor 76*:319–25.

Paladino, F. V., and King, J. R. (1979) Energetic cost of terrestrial locomotion: biped and quadruped runners compared. *Rev. Can. Biol. 38*:321–3.

Parrot, C. (1894) Ueber die Grössenverhältnisse des Herzens bei Vögeln. *Zool. Jahr. (Syst.) 7*:496–522.

Pasquis, P., and Dejours, P. (1965) Consommation maximale d'oxygene chez le rat blanc et le cobaye. *J. Physiol. (Paris) 57*:670.

Pasquis, A., Lacaisse, A., and Dejours, P. (1970) Maximal oxygen uptake in four species of small mammals. *Resp. Physiol. 9*:298–309.

Pedley, T. J. (ed.) (1977) *Scale Effects in Animal Locomotion.* London: Academic Press. 545 pp.

Pennycuick, C. J. (1969) The mechanics of bird migration. *Ibis 111*:525–56.

– (1975) On the running of the gnu (*Connochaetes taurinus*) and other animals. *J. Exp. Biol. 63*:775–99.

Peters, R. H. (1983) *The Ecological Implications of Body Size.* Cambridge University Press. 324 pp.

Pough, F. H. (1977) The relationship of blood oxygen affinity to body size in lizards. *Comp. Biochem. Physiol. 57A*:435–41.

Prange, H. D., Anderson, J. F., and Rahn, H. (1979) Scaling of skeletal mass to body mass in birds and mammals. *Amer. Nat. 113*:103–22.

Price, J. W. (1931). Growth and gill development in the small-mouthed black bass, *Micropterus dolomieu* Lacepede. Franz Theodore Stone Laboratory Contribution No. 4. Columbus: Ohio State University Press. 46 pp.

Pritchard, A. W., Florey, E., and Martin, A. W. (1958) Relationship between metabolic rate and body size in an elasmobranch (*Squalus suckleyi*) and in a teleost (*Ophiodon elongatus*). *J. Mar. Res. 17*:403–11.

Prothero, J. (1979) Heart weight as a function of body weight in mammals. *Growth 43*:139–50.

– (1980) Scaling of blood parameters in mammals. *Comp. Biochem. Physiol. 67A*:649–57.

Rahn, H., and Ar, A. (1974) The avian egg: incubation time and water loss. *Condor 76*:147–52.

Rahn, H., Paganelli, C. V., and Ar, A. (1975) Relation of avian egg weight to body weight. *Auk 92*:750–65

Ralph, R., and Everson, I. (1968) The respiratory metabolism of some Antarctic fish. *Comp. Biochem. Physiol. 27*:299–307.

Rashevsky, N. (1960) *Mathematical Biophysics. Physico-mathematical Foundations of Biology, Vols. 1 and 2* (ed. 3). New York: Dover Publications.

Reynolds, W. W., and Karlotski, W. J. (1977) The allometric relationship of skeleton weight to body weight in teleost fishes: a preliminary comparison with birds and mammals. *Copeia 1977*:160–3.

Riggs, A. (1960) The nature and significance of the Bohr effect in mammalian hemoglobins. *J. Gen. Physiol. 43*:737–52.

Romer, A. S. (1966) *Vertebrate Paleontology* (ed. 3). University of Chicago Press. 468 pp.

Rouse, H. (1961) *Fluid Mechanics for Hydraulic Engineers*. New York: Dover. 422 pp.

Royal Society, Symbols Committee (1975) *Quantities, Units, and Symbols* (ed. 2). London: Royal Society. 54 pp.

Rubner, M. (1883) Ueber den Einfluss der Körpergrösse auf Stoffund Kraftwechsel. *Z. Biol. 19*:535–62.

Sacher, G. A. (1959) Relation of lifespan to brain weight and body weight in mammals. In *Ciba Foundation Colloquium on Aging, Vol. 1* (G. E. W. Wolstenholme, ed.), pp. 115–41.

Saltin, B., and Åstrand, P.-O. (1967) Maximal oxygen uptake in athletes. *J. Appl. Physiol. 23*:353–8.

Sarrus et Rameaux (1838–9) Rapport sur un mémoire adressé a l'Académie Royale de Médecine. Commissaires Robiquet et Thillaye, rapporteurs. *Bull. Acad. Roy. Med. (Paris) 3*:1094–100.

Saunders, R. L. (1963) Respiration of the Atlantic cod. *J. Fish. Res. Bd. Canada 20*:373–86.

Scheid, P., and Piiper, J. (1972) Cross-current gas exchange in avian lungs: effects of reversed parabronchial air flow in ducks. *Resp. Physiol. 16*:304–12.

Schmidt-Nielsen, K. (1951) Tissue respiration and body size. *Science 114*:306.

– (1964) *Desert Animals. Physiological Problems of Heat and Water*. Oxford: Clarendon Press. 277 pp.

– (1970) Energy metabolism, body size, and problems of scaling. *Fed. Proc. 29*:1524–32.

– (1972) *How Animals Work*. Cambridge University Press. 114 pp.

– (1975a) Scaling in biology: the consequences of size. *J. Exp. Zool. 194*:287–308.

– (1975b) Recent advances in avian respiration. *Symp. Zool. Soc. Lond. 35*:33–47.

– (1983) *Animal Physiology. Adaptation and Environment* (ed. 3). Cambridge University Press. 619 pp.

Schmidt-Nielsen, K., Dawson, T. J., and Crawford, E. C., Jr. (1966) Temperature regulation in the echidna (*Tachyglossus acculeatus*). *J. Cell. Physiol. 67*:63–71.

Schmidt-Nielsen, K., and Pennycuik, P. (1961) Capillary density in mammals in relation to body size and oxygen consumption. *Amer. J. Physiol. 200*:746–50.

Schmitt, W. L. (1965) *Crustaceans*. Ann Arbor: University of Michigan Press. 204 pp.

Scholander, P. F., Hock, R., Walters, V., Johnson, F., and Irving, L. (1950a) Heat regulation in some arctic and tropical mammals and birds. *Biol. Bull. 99*:237–58.

Scholander, P. F., Walters, V., Hock, R., and Irving, L. (1950b) Body insulation of some arctic and tropical mammals and birds. *Biol. Bull. 99*:225–36.

Segrem, N. P., and Hart, J. S. (1967) Oxygen supply and performance in *Peromyscus*. Metabolic and circulatory responses to exercise. *Can. J. Physiol. Pharmacol. 45*:531–41.

Sheets, R. G., Linder, R. L., and Dahlgren, R. B. (1971) Burrow systems of prairie dogs in South Dakota. *J. Mammal.* 52:451–3.

Smith, A. H. (1976) Physiological changes associated with long-term increases in acceleration. In *COSPAR: Life Sciences and Space Research 14* (P. H. A. Sneath, ed.), pp. 91–100. Berlin: Akademie-Verlag.

– (1978) The roles of body mass and gravity in determining the energy requirements of homoiotherms. In *COSPAR: Life Sciences and Space Research 16* (R. Holmquist and A. C. Stickland, eds.), pp. 83–8. Oxford: Pergamon Press.

Smith, A. H., and Pace, N. (1971) Differential component and organ size relationship among whales. *Environ. Physiol.* 1:122–36.

Smith, R. E. (1956) Quantitative relations between liver mitochondria metabolism and total body weight in mammals. *Ann. N.Y. Acad. Sci.* 62:403–22.

Smith, R. J. (1980) Rethinking allometry. *J. Theor. Biol.* 87:97–111.

Snell, O. (1891) Das Gewicht des Gehirnes und des Hirnmantels der Säugethiere in Beziehung zu deren geistigen Fähigkeiten. *Sitzungsberichte der Gesellschaft für Morphologie und Physiologie in München* 7:90–4.

Spells, K. E. (1969) Comparative studies in lung mechanics based on a survey of literature data. *Resp. Physiol.* 8:37–57.

Spotila, J. R., Lommen, P. W., Bakken, G. S., and Gates, D. M. (1973) A mathematical model for body temperatures of large reptiles: implications for dinosaur ecology. *Amer. Nat.* 107:391–404.

Stahl, W. R. (1965) Organ weights in primates and other mammals. *Science* 150:1039–42.

– (1967) Scaling of respiratory variables in mammals. *J. Appl. Physiol.* 22:453–60.

Stolpe, M., and Zimmer, K. (1939) Der Schwirrflug des Kolibri im Zeitlupenfilm. *J. Ornithologie* 87:136–55.

Szarski, H. (1964) The structure of respiratory organs in relation to body size in amphibia. *Evolution* 18:118–26.

Szarski, H., and Czopek, G. (1966) Erythrocyte diameter in some amphibians and reptiles. *Bull. Acad. Polonaise Sci. (Cl. II)* 14:433–7.

Taylor, C. R., Caldwell, S. L., and Rowntree, V. J. (1972) Running up and down hills: some consequences of size. *Science* 178:1096–7.

Taylor, C. R., Heglund, N. C., McMahon, T. A., and Looney, T. R. (1980) Energetic cost of generating muscular force during running. *J. Exp. Biol.* 86:9–18.

Taylor, C. R., Heglund, N. C., and Maloiy, G. M. O. (1982) Energetics and mechanics of terrestrial locomotion. I. Metabolic energy consumption as a function of speed and body size in birds and mammals. *J. Exp. Biol.* 97:1–21.

Taylor, C. R., Maloiy, G. M. O., Weibel, E. R., Langman, V. A., Kamau, J. M. Z., Seeherman, H. J., and Heglund, N. C. (1981) Design of the mammalian respiratory system. III. Scaling maximum aerobic capacity to body mass: wild and domestic mammals. *Resp. Physiol.* 44:25–37.

Taylor, C. R., and Rowntree, V. J. (1973) Running on two or on four legs: Which consumes more energy? *Science* 179:186–7.

Taylor, C. R., Schmidt-Nielsen, K., and Raab, J. L. (1970) Scaling of energetic cost of running to body size in mammals. *Amer. J. Physiol.* 219:1104–7.

Taylor, C. R., and Weibel, E. R. (1981) Design of the mammalian respiratory system. I. Problem and strategy. *Resp. Physiol.* 44:1–10.

Tazawa, H., Mochizuki, M., and Piiper, J. (1979) Blood oxygen dissociation curve of the frogs *Rana catesbeiana* and *Rana brevipoda*. *J. Comp. Physiol.* 129:111–14.

Teissier, G. (1939) Biometrie de la cellule. *Tabulae Biologicae 19(Pt. 1)*:1–64.

Tenney, S. M., and Bartlett, D., Jr. (1967) Comparative quantitative morphology of the mammalian lung: trachea. *Resp. Physiol.* 3:130–5.

Tenney, S. M., and Morrison, D. H. (1967) Tissue gas tensions in small wild mammals. *Resp. Physiol. 3*:160–5.

Tenney, S. M., and Remmers, J. E. (1963) Comparative quantitative morphology of the mammalian lung: diffusing area. *Nature 197*:54–6.

Tenney, S. M., and Tenney, J. B. (1970) Quantitative morphology of cold-blooded lungs: amphibia and reptilia. *Resp. Physiol. 9*:197–215.

Terroine, É. F., and Roche, J. (1925) La respiration des tissus. I. Production calorique des Homéothermes et intensité de la respiration *in vitro* des tissus homologues. *Arch. Intern. Physiol. 24*:356–99.

Thomas, R. D. K., and Olson, E. C. (eds.) (1980) *A Cold Look at the Warm-Blooded Dinosaurs.* AAAS Selected Symposium 28. Boulder, CO: Westview Press. 514 pp.

Thomas, S. P. (1975) Metabolism during flight in two species of bats, *Phyllostomus hastatus* and *Pteropus gouldii. J. Exp. Biol. 63*:273–93.

– (1981) Ventilation and oxygen extraction in the bat *Pteropus gouldii* during rest and steady flight. *J. Exp. Biol. 94*:231–50.

Thompson, D. W. (1961) *On Growth and Form* (an abridged edition edited by J. T. Bonner). Cambridge University Press. 346 pp.

Tucker, V. A. (1968) Respiratory exchange and evaporative water loss in the flying budgerigar. *J. Exp. Biol. 48*:67–87.

– (1973) Bird metabolism during flight: evaluation of a theory. *J. Exp. Biol. 58*:689–709.

Tucker, V. A., and Schmidt-Koenig, K. (1971) Flight speeds of birds in relation to energetics and wind directions. *Auk 88*:97–107.

Ulrich, S. P. H., and Bartels, H. (1963) Über die Atmungsfunktion des Blutes von Spitzmäusen, weissen Mäusen und syrischen Goldhamstern. *Pflügers Arch. 277*:150–65.

Ultsch, G. R. (1973) A theoretical and experimental investigation of the relationships between metabolic rate, body size, and oxygen exchange capacity. *Resp. Physiol. 18*:143–60.

– (1974) Gas exchange and metabolism in the Sirenidae (Amphibia: Caudata)—I. Oxygen consumption of submerged sirenids as a function of body size and respiratory surface area. *Comp. Biochem. Physiol. 47A*:485–98.

Umminger, B. L. (1975) Body size and whole blood sugar concentrations in mammals. *Comp. Biochem. Physiol. 50A*:455–8.

Ursin, E. (1967) A mathematical model of some aspects of fish growth, respiration, and mortality. *J. Fish. Res. Bd. Canada 24*:2355–453.

Vogel, P. (1976) Energy consumption of European and African shrews. *Acta Theriol. 21*:195–206.

Voit, E. (1930) Die Messung und Berechnung der Oberfläche von Mensch und Tier. *Z. Biol. 90*:237–59.

Wainwright, S. A., Biggs, W. D., Currey, J. D., and Gosline, J. M. (1976) *Mechanical Design in Organisms.* London: Edward Arnold. 424 pp.

Wangensteen, O. D., Wilson, D., and Rahn, H. (1971) Diffusion of gases across the shell of the hen's egg. *Resp. Physiol. 11*:16–30.

Webb, P. W. (1978) Hydrodynamics: nonscombroid fish. In *Fish Physiology, Vol. VII* (W. S. Hoar and D. J. Randall, eds.), pp. 189–237. New York: Academic Press.

Weibel, E. R. (1972) Morphometric estimation of pulmonary diffusion capacity. V. Comparative morphometry of alveolar lungs. *Resp. Physiol. 14*:26–43.

Weibel, E. R., Burri, P. H., and Claassen, H. (1971) The gas exchange apparatus of the smallest mammal: *Suncus etruscus. Experientia 27*:724.

Weibel, E. R., Taylor, C. R., Gehr, P., Hoppeler, H., Mathieu, O., and Maloiy, G. M. O. (1981) Design of the mammalian respiratory system. IX. Functional and structural limits for oxygen flow. *Resp. Physiol. 44*:151–64.

Weis-Fogh, T. (1964) Diffusion in insect wing muscle, the most active tissue known. *J. Exp. Biol. 41*:229–56.

- (1972) Energetics of hovering flight in hummingbirds and in *Drosophila. J. Exp. Biol. 56*:79–104.

- (1973) Quick estimates of flight fitness in hovering animals, including novel mechanisms for lift production. *J. Exp. Biol. 59*:169–230.

- (1977) Dimensional analysis of hovering flight. In *Scale Effects in Animal Locomotion* (T. J. Pedley, ed.), pp. 405–20. London: Academic Press.

Wilkie, D. R. (1959) The work output of animals: flight by birds and by man-power. *Nature 183*:1515–16.

Wohlschlag, D. E. (1963) An antarctic fish with unusually low metabolism. *Ecology 44*:557–64.

Wood, S. C., Johansen, K., Glass, M. L., and Maloiy, G. M. O. (1978) Aerobic metabolism of the lizard *Varanus exanthematicus:* effects of activity, temperature, and size. *J. Comp. Physiol. 127*:331–6.

Wyatt, G. R. (1967) The biochemistry of sugars and polysaccharides in insects. In *Advances in Insect Physiology, Vol. 4* (J. W. L. Beament, J. E. Treherne, and V. B. Wigglesworth, eds.), pp. 287–360. London: Academic Press.

Young, D. R., Mosher, R., Erve, P., and Spector, H. (1959) Energy metabolism and gas exchange during treadmill running in dogs. *J. Appl. Physiol. 14*:834–8.

Yousef, M. K., and Johnson, H. D. (1975) Thyroid activity in desert rodents: a mechanism for lowered metabolic rate. *Amer. J. Physiol. 229*:427–31.

Zar, J. H. (1968) Calculation and miscalculation of the allometric equation as a model in biological data. *BioScience 18*:1118–20.

Zerbe, G. O., Archer, P. G., Banchero, N., and Lechner, A. J. (1982) On comparing regression lines with unequal slopes. *Amer. J. Physiol. 242*:R178–80.

Zeuthen, E. (1947) Body size and metabolic rate in the animal kingdom. *Compt. Rend. Lab. Carlsberg, Ser. Chim. 26*:17–165.

- (1953) Oxygen uptake as related to body size in organisms. *Quart. Rev. Biol. 28*:1–12.

Index